T0296757

PARKINSON'S DISEASE THERAPEUTICS

PARKINSON'S DISEASE THERAPEUTICS

Emphasis on Nanotechnological Advances

MAGISETTY OBULESU
Scientist,
ATG Laboratories, Pune
Certified by Stanford University for Scientific Writing

Academic Press is an imprint of Elsevier
125 London Wall, London EC2Y 5AS, United Kingdom
525 B Street, Suite 1650, San Diego, CA 92101, United States
50 Hampshire Street, 5th Floor, Cambridge, MA 02139, United States
The Boulevard, Langford Lane, Kidlington, Oxford OX5 1GB, United Kingdom

Notices

Knowledge and best practice in this field are constantly changing. As new research and experience broaden our understanding, changes in research methods, professional practices, or medical treatment may become necessary.

Practitioners and researchers must always rely on their own experience and knowledge in evaluating and using any information, methods, compounds, or experiments described herein. In using such information or methods they should be mindful of their own safety and the safety of others, including parties for whom they have a professional responsibility.

To the fullest extent of the law, neither the Publisher nor the authors, contributors, or editors, assume any liability for any injury and/or damage to persons or property as a matter of products liability, negligence or otherwise, or from any use or operation of any methods, products, instructions, or ideas contained in the material herein.

Library of Congress Cataloging-in-Publication Data
A catalog record for this book is available from the Library of Congress

British Library Cataloguing-in-Publication Data
A catalogue record for this book is available from the British Library

ISBN: 978-0-12-819882-7

For information on all Academic Press publications visit our website at https://www.elsevier.com/books-and-journals

Publisher: Nikki Levy
Acquisitions Editor: Natalie Farra
Editorial Project Manager: Tracy I. Tufaga
Production Project Manager: Bharatwaj Varatharajan
Cover Designer: Greg Harris

Typeset by TNQ Technologies

This work is dedicated to my beloved parents (Magisetty Jagan Mohan and Magisetty Saraswathi), wife, and all my family members (sisters, daughter, and son). I had a praiseworthy mental support during this project and throughout my life.

Contents

Biography *xi*
Acknowledgments *xiii*

1. Introduction: Parkinson's disease pathology and therapeutics **1**
1. Introduction 1
2. Micro-ribonucleic acid 1
3. Drugs 2
4. Viral vectors 3
5. Conclusions and future perspectives 6
References 6
Further reading 11

2. Biomaterials: a boon or a bane in the treatment of Parkinson's disease **13**
1. Introduction 13
2. Oxidative stress 13
3. Nanoparticles 14
4. Neurturin 15
5. Delivery systems for natural compounds 15
6. Diagnosis 16
7. Silver nanoparticles 17
8. Immunotherapy 17
9. Shortcomings of nanotechnology 17
10. Conclusions and future perspectives 18
References 18
Further reading 23

3. Blood—brain barrier targeted nanoparticles **25**
1. Introduction 25
2. Blood—brain barrier impairment 26
3. Blood—brain barrier spanning of nanoparticles 28
4. Parkinson's disease 28
5. Nanoparticles for Parkinson's disease 29
6. Gold nanoparticles 29
7. Quercetin nanoparticles 30

8. Conclusions and future perspectives 33
References 33
Further reading 38

4. Natural compounds in the treatment of Parkinson's disease 39
1. Introduction 39
2. Plant-derived compounds 40
3. Amelioration of antioxidant enzymes 43
4. Conclusions and future perspectives 45
References 45

5. Curcumin: a promising therapeutic in Parkinson's disease treatment 51
1. Introduction 51
2. Antioxidant activity 51
3. 6-Hydroxydopamine and MPTP PD models 52
4. α-Synuclein targeting 52
5. Curcumin derivatives 53
6. Bioavailability 53
7. Heat shock proteins 53
8. Antiapoptosis 54
9. Metal chelation 54
10. Delivery systems for curcumin 55
11. Conclusions and future perspectives 56
References 59

6. Redox nanoparticles: the corner stones in the development of Parkinson's disease therapeutics 65
1. Introduction 65
2. Parkinson's disease diagnosis 66
3. Coenzyme Q10 66
4. Conclusions and future perspectives 69
References 70

7. Liposomal maneuvers against Parkinson's disease 75
1. Introduction 75
2. Blood—Brain Barrier 75
3. Liposomes 76
4. Surface functionalization in brain targeting 77
5. Demerits of liposomes 83

6. Conclusions and future perspectives 83
References 83
Further reading 88

8. Nasal delivery nanoparticles: An insight into novel Parkinson's disease therapeutics 89
1. Introduction 89
2. Nasal delivery 89
3. Challenges in nasal delivery 91
4. Demerits of nose to brain delivery 95
5. Conclusions and future perspectives 96
References 96
Further reading 101

9. α-Synuclein-targeted nanoparticles 103
1. Introduction 103
2. 1-Methyl-4-phenyl-1,2,3,6-tetrahydropyridine treatment 104
3. Fibrillation 105
4. Early diagnosis 105
5. Conclusions and future perspectives 109
References 109
Further reading 114

10. Pros and cons of Parkinson's disease therapeutics 115
1. Introduction 115
2. α-Synuclein 115
3. *Caenorhabditis elegans*: a potential animal model for Parkinson's disease 116
4. Blood–brain barrier 116
5. Stem cell therapy 117
6. Influence of gut microbiota on neurodegenerative diseases 117
7. Conclusions and future perspectives 121
References 122

Index *129*

Biography

M. Obulesu, MSc (PhD), is a Scientist in ATG Laboratories, Pune, India. He has 19 years of research and teaching experience. His research areas are multifarious, which include food science, pathology of neurodegenerative diseases such as Alzheimer's disease, and designing polymer-based biomaterials (design of hydrogels, etc.). He did Alzheimer's disease research and developed an aluminum-induced neurotoxicity rabbit model. His present research focuses on development of redox-active injectable hydrogels of polyion complex. His research area also includes development of metal chelators to overcome metal-induced toxicity. His books, three monographs entitled "Alzheimer's Disease Theranostics (Elsevier (First Edition))," "Parkinson's Disease -Therapeutics: Emphasis on Nanotechnological Advances" (Elsevier (First Edition)), "Turmeric and Curcumin for Neurodegenerative Diseases (Elsevier (First Edition))," and an editorial collection "Phytomedicine and Alzheimer's Disease (Taylor and Francis (First Edition))," are accepted currently. He is the first and corresponding author for majority of his articles. He secured a certificate from Stanford University for the Scientific Writing course. He is on editorial board of a few pathology journals such as Journal of Medical Laboratory and Diagnosis, Journal of Medical and Surgical Pathology, Annals of Retrovirals and Antiretrovirals, Kenkyu Journal of Medical Science and Clinical Research, SciFed Oncology and Cancer Research Journal, and Journal of Cancer and Cure. He is also on editorial board of nanotechnology journals such as SciFed Nanotech Research Letters, SciFed Drug Delivery Research, Current Updates in Nanotechnology, and Journal of Nanotechnology and Materials Science.

Acknowledgments

I sincerely acknowledge the endeavors of my ever-memorable teacher Mr. Nageswara Rao, Director, Apex Institute of English, Guntur, Andhra Pradesh, India. His worth-commending didactic skills made my English language skills reach culminating points. I also acknowledge all my teachers and mentors who significantly nurtured my research and writing skills.

CHAPTER 1

Introduction: Parkinson's disease pathology and therapeutics

1. Introduction

1.1 Pathology

Parkinson's disease (PD) has been considered the second most common neurodegenerative disease after Alzheimer's disease (AD) since its identification by the English physician James Parkinson in 1817 (Pringsheim et al., 2014; Gao et al., 2019). Reactive oxygen species (ROS) generation, oxidative stress, apoptosis, genetic mutations, and metal intoxication have long been implicated in the etiology of neurodegenerative diseases such as AD and PD (Obulesu et al., 2009, Obulesu and Muralidhara Rao, 2010, 2011a,b, Obulesu and Jhansilakshmi, 2014a,b, 2016, Obulesu, 2018). PD is the most common neurodegenerative disorder in the elderly. The prevalence of PD is about 1 in every 100 people over 65 years age (Siddiqi et al., 2018). The prominent pathological characteristics involved in PD are formation of amyloid protein fibrils, deterioration of neurotransmitter, and dopamine (DA) (Mohammadi and Nikkah, 2017). As DA actively plays a role in electric signal transmission, which is important in normal physical movement, its low levels result in bradykinesia, rigidity, and resting tremor (Ramanathan et al., 2018). In addition, pathological Lewy bodies are initially observed in brain stem and olfactory bulb, which further extend to cortex and nigra (Ramanathan et al., 2018). PD patient's brain showed low levels of astroglial cells because of the downstream processing in brain (Orr et al., 2002; Mena, 2008; Obeso et al., 2010, Ramanathan et al., 2018).

2. Micro-ribonucleic acid

Micro-ribonucleic acid (miRNA), which is small noncoding RNA, plays a pivotal role in posttranscriptional gene silencing. Abnormal expressions of miRNA have long been implicated in neurodegenerative disorders such as

Parkinson's Disease Therapeutics
ISBN 978-0-12-819882-7
https://doi.org/10.1016/B978-0-12-819882-7.00001-5

PD (Hebert and De Strooper, 2009, Titze de Almeida et al., 2018). Antisense single-stranded oligos (AntimiRs) were found to reverse rotenone-induced damage in dopaminergic SH-SY5Y cells (Horst et al., 2017, Titze de Almeida et al., 2018). In line with this, intracerebroventricular injection of AntimiRs into rat brain probably protects striatum, which is usually involved in PD pathology (Titze de Almeida et al., 2018).

3. Drugs

Levodopa has been extensively used drug in the treatment of PD. However, it poses a few biological riddles such as gastric pH; intervention of proteins with bioavailability of drugs limits the success of drugs (Ramanathan et al., 2018). Rotigotine exhibits neuroprotection by improving motor symptoms in PD (Fig. 1.1) (Stockwell et al., 2010; Yu et al., 2015; Yan et al., 2018a,b). Rotigotine patch comprises drug in elastic vessels and released in silicon matrix via ionic gradients resulting in equal distribution in the epidermis (Johnston et al., 2005; Katzenschlager et al., 2005; Nyholm et al., 2005; Bennet and Piercy, 1999, Ramanathan et al., 2018). Despite the lipophilicity of the drugs, their efflux has frequently been observed because of the occurrence of multispecific organic anion transporter (MOAT), P-glycoprotein (Pgp), and multidrug resistance proteins (MRPs) (Hans and Lowman, 2002; Persidsky et al., 2006, Ramanathan et al., 2018).

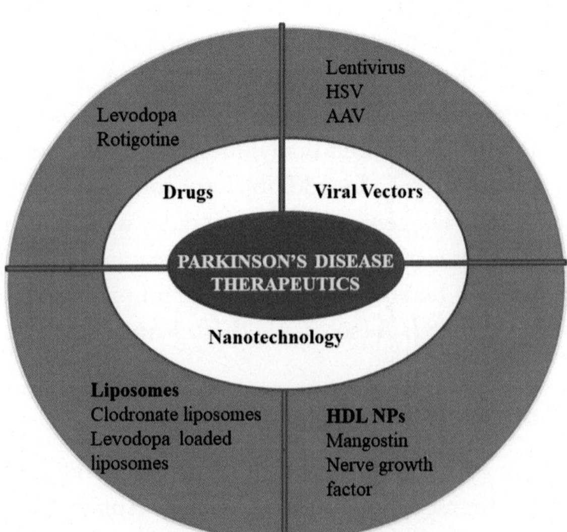

Figure 1.1 Multifarious therapeutic avenues targeted against Parkinson's Disease.

Intranasal introduction of drugs has been a noninvasive method of drug delivery to the brain to overcome blood–brain barrier (BBB) (Ali et al., 2010; Lalani et al., 2014; Yan et al., 2018a,b). Because nasal delivery has a few advantages such as decreased drug delivery to nontarget tissues, averting hepatic first-pass metabolism, drugs such as rasagiline, alginate, tarenflurbil, and piperine are introduced through this route (Haque et al., 2014; Elnaggar et al., 2015; Muntimadugu et al., 2016; Mittal et al., 2016; Yan et al., 2018a,b).

4. Viral vectors

Viral vectors with the capability to infect cells with nucleic acids have also been used as amenable vectors to overcome pathology of several diseases such as PD (Dong et al., 2018). Surprisingly, they offer enhanced transfection efficacy of 80% (Perez-Martinez et al., 2012). A few viral vectors employed for brain delivery include lentivirus, herpes simplex virus (HSV), adenovirus, and adeno-associated virus (AAV) (Fig. 1.1). Despite immunogenicity concerns, AAV exhibited remarkably improved brain delivery and significantly better safety profiles in humans (Mingozzi and High, 2013, Dong et al., 2018). Despite the progress of new viral vectors in the treatment of brain diseases, a few studies involved direct injection of viral vectors into the brain (Natarajan et al., 2017; Tanabe et al., 2017, Dong et al., 2018). Plethora of hitherto employed AAV serotypes exhibited substantial ability to overcome BBB and localize in cells of central nervous system (Zhang et al., 2011; Ahmed et al., 2013; Vagner et al., 2016, Dong et al., 2018). However, a few challenges such as expensive production methods, hurdles in manufacturing, and safety impede their success (Hollon, 2000; Check, 2005, Dong et al., 2018).

4.1 Nanotechnology

Several lines of evidence indicate that nanoparticles (NPs) with small size are effective in achieving targeted delivery of loaded therapeutics, enhanced systemic circulation, and enhanced permeation and retention capabilities (Huang and Liu, 2011, Dong et al., 2018). Therefore, brain delivery of NPs with the aid of manifold transporters and ligands has been in practice for about three decades (Georgieva et al., 2014, Dong et al., 2018). NPs have extensively been used in tissue engineering and drug delivery (Siddiqi et al., 2018). They play an essential role in diagnosis and treatment of neurodegenerative diseases (Wilson, 2009; Lauzon et al., 2015, Siddiqi et al., 2018).

A wide range of NPs in use for drug delivery currently are magnetic NPs, metal NPs, and quantum dots (The Royal society and The royal academy of Engineering, 2018; Neuwelt et al., 2011, Ramanathan et al., 2018). Silver and gold NPs have been widely used in disease diagnosis, therapy, and sensor development (Sharma et al., 2009, Lara et al., 2011, Husen and Siddiqi, 2014, Siddiqi and Husen, 2016, Siddiqi and Husen, 2017, Husen, 2017, Siddiqi et al., 2018). Although a few peptides and peptide NPs are found to be effective in the treatment of PD currently, yet a significant impetus in novel technologies has been warranted (Anna, 2015, Siddiqi et al., 2018).

4.2 Multifarious nanoparticles and surface functionalization

Panoply of brain delivery systems include polymeric nanomaterials, liposomes, carbon nanotubes (CNTs), and surface-functionalized NPs which are superior in performance compared to traditional delivery systems (Davson and Segal, 1996, Begley and Kreuter, 1999, Weksler et al., 2005, Juillerat-Jeanneret, 2008, Fonseca-Santos et al., 2015, Bolognesi, 2017, Sikkandhar et al., 2017, Sridhar et al., 2017, Ramanathan et al., 2018). While conjugation of polysorbate 80 with NPs facilitates them to pass through BBB, a few alterations in preparation aid them to evade enzymatic degradation/immune function (Wilson, 2009; Locatelli and Franchini, 2012, Siddiqi et al., 2018). Although a possible mechanism of NP localization has been identified as a receptor-mediated transcytosis, yet the targeted NP delivery into brain is an ambiguous issue until now (Candela et al., 2008, Dong et al., 2018). Additionally, a few studies also emphasized that NPs localization followed by increased drug absorption in brain is nonspecific (Chen et al., 2008; Xu et al., 2009, Masserini, 2013, Dong et al., 2018).

4.3 Nanoparticles in drug and gene delivery

Neurotensin (NTS) polyplex delivery system for targeted genes was found to ameliorate PD and cancer symptoms significantly (Barradas et al., 2018). Using this delivery system gene expressing NTS receptor type 1 (NTSR1) has been delivered into brain (Martinez-Fong et al., 2009, Barradas et al., 2018). Intracranial injection of NTS polyplex exhibited enhanced and prolonged expression of neurotrophic factor genes in DA neurons of hemiparkinsonian rats (Gonzalez-Barrios et al., 2006; Hernandez-Chan et al., 2015; Razgado-Hernandez et al., 2015, Barradas et al., 2018).

Lactoferrin-modified NPs have been used to deliver rotigotine to the brain effectively and ameliorate PD symptoms (Yan et al., 2018a,b). Lactoferrin shows profound ability to span BBB; therefore, it has been extensively used in drug delivery to the brain (Kumar et al., 2015; Kumari et al., 2017; Yan et al., 2018a,b). As evaluation of DA levels plays an essential role in PD diagnosis, nanotechnology has been used to detect the same (Yan et al., 2018a,b). In line with this, Ag—Fe_3O_4 NPs were designed and used to successfully estimate DA levels with limit of detection of 10 μM (Yan et al., 2018a,b). Recently designed gold and TiO_2-associated nanotubes effectively detect α-synuclein by photoelectrochemical immune sensors (An et al., 2010, Siddiqi et al., 2018). Furthermore, α-synuclein tethered polybutylcyanoacrylate NPs aid α-synuclein clearance (Hasadsri et al., 2009; Klyachko et al., 2014, Siddiqi et al., 2018). CNTs discovered by Iijima in 1991 were extensively used in the treatment of PD (Kakkar and Dahiya, 2015, Siddiqi et al., 2018). They have been preferentially used as drug delivery vehicles (Malarkey and Parpura, 2007, Siddiqi et al., 2018).

4.4 High-density lipoprotein nanoparticles

Recent study has shown that apolipoprotein E (ApoE)—associated high-density lipoprotein (HDL) NPs encapsulated with a polyphenolic compound, α-mangostin, hindered Aβ oligomer formation (Song et al., 2016, Dong et al., 2018). In another study, nerve growth factor (NGF)—loaded HDL NPs synthesized using natural lipids showed better bioactivity of neurite outgrowth in pheochromocytoma (PC12) cells and enhanced circulation of NGF in mice (Fig. 1.1) (Prathipati et al., 2016; Zhu and Dong, 2017, Dong et al., 2018).

4.5 Liposomes

They are readily biodegradable lipophilic molecules which do not induce any immunogenic reactions, thrombosis, and cancer (Gregoriadis, 2008, Siddiqi et al., 2018). A few advantages of liposomes include their ability to load manifold components, span BBB, safeguard from enzymatic degradation, and escape from reticuloendothelial system removal (Siddiqi et al., 2018). Additionally, the half-life of liposomes can feasibly be improved by conjugating surface with polyethylene glycol (Spuch and Navarro, 2011, Siddiqi et al., 2018). PD-targeted liposomes include clodronate liposomes (Yan et al., 2018a,b), levodopa-loaded liposomes co-loaded with antioxidants such as curcumin, ascorbic acid, and superoxide dismutase (Garcia-Esteban et al., 2018).

4.6 Demerits of nanotechnology

Wealth of studies has emphasized that titanium dioxide (TiO_2) NPs induce PD-like symptoms in zebrafish via generation of ROS, thus resulting in cell death in hypothalamus region of the brain (Hu et al., 2017, Siddiqi et al., 2018). In addition, nanotechnology has not been so effective in producing substantial therapeutic arsenals, although it is effective for numerous other diseases.

5. Conclusions and future perspectives

PD has been the most deleterious neurodegenerative disorder with complex etiology after AD. Manifold drugs, viral vectors, drug delivery systems, and miRNAs were used to overcome PD until now. However, BBB has been a potential barrier and hurdle in achieving the successful therapeutic localization in brain. BBB has been found to deteriorate drug concentration, thus significantly decreasing therapeutic effect of corresponding drug (Vashist et al., 2018, Siddiqi et al., 2018).

Despite the discovery of multifarious therapeutics, nanovectors and their remarkable progress in PD treatment, as well as a few hitherto unresolved issues impede their success (Barradas et al., 2018]. Furthermore, a wide range of ligands have been extensively used in NPs' surface functionalization to accomplish brain targeting, although it is insufficient to achieve substantial therapeutic efficacy. Therefore, to overcome the above-mentioned major issues in drug delivery to the brain, a multidimensional approach that incorporates medicine, pharmacy, and biomaterials is essential to develop a substantial therapeutic arsenal.

References

Ahmed, S.S., Li, H., Cao, C., Sikoglu, E.M., Denninger, A.R., Su, Q., et al., 2013. A single intravenous rAAV injection as late as P20 achieves efficacious and sustained CNS gene therapy in Canavan mice. Mol. Ther. 21, 2136–2147.

Ali, J., Ali, M., Baboota, S., Ali, J., 2010. Potential of nanoparticulate drug delivery systems by intranasal administration. Curr. Pharmaceut. Des. 16, 1644–1653.

Anna, P.N., 2015. Nanotechnology and its applications in medicine. Med. Chem. 5, 2.

An, Y., Tang, L., Jiang, X., Chen, H., Yang, M., Jin, L., et al., 2010. Photoelectrochemical immunosensor based on Au-doped TiO2 nanotube arrays for the detection of α-synuclein. Chemistry 16, 14439–14446.

Bolognesi, M.L., 2017. From Imaging Agents to Theranostic Drugs in Alzheimer's Disease. Reference Module in Chemistry, Molecular Sciences and Chemical Engineering. Elsevier, New York, pp. 74–106, 2017.

Begley, D.J., Kreuter, J., 1999. Do ultra-low frequency (ULF) magnetic fields affect the blood brain barrier? In: Holick, M.F., Jung, E.G. (Eds.), Biologic Effects of Light 1998. Springer US, Boston, MA, pp. 297–301, 1999.

Bennett, J.P., Piercey, M.F., 1999. Pramipexole – a new dopamine agonist for the treatment of Parkinson's disease. J. Neurol. Sci. 163, 25–31.

Check, E., 2005. Gene therapy put on hold as third child develops cancer. Nature 433, 561.

Candela, P., Gosselet, F., Miller, F., Buee-Scherrer, V., Torpier, G., Cecchelli, R., et al., 2008. Physiological pathway for low-density lipoproteins across the blood-brain barrier: transcytosis through brain capillary endothelial cells in vitro. Endothelium 15, 254–264.

Chen, H., Kim, S., Li, L., Wang, S., Park, K., Cheng, J.X., 2008. Release of hydrophobic molecules from polymer micelles into cell membranes revealed by Forster resonance energy transfer imaging. Proc. Natl. Acad. Sci. U.S.A. 105, 6596–6601.

Davson, H., Segal, M.B., 1996. Physiology of the CSF and Blood-Brain Barriers. CRC Press, Boca Raton, p. 11.

Dong, X., 2018. Current Strategies for Brain Drug Delivery. Theranostics 8, 1481–1493.

Elnaggar, Y.S., Etman, S.M., Abdelmonsif, D.A., Abdallah, O.Y., 2015. Intranasal piperine-loaded Chitosan nanoparticles as brain-targeted therapy in Alzheimer's disease: optimization, biological efficacy, and potential toxicity. J. Pharm. Sci. 104, 3544–3556.

Fonseca-Santos, B., Gremiao, M.P.D., Chorilli, M., 2015. Nanotechnology-based drug delivery systems for the treatment of Alzheimer's disease. Int. J. Nanomed. 10, 4981–5003.

Gao, H., Zhao, Z., He, Z., Wang, H., Liu, M., Hu, Z., et al., 2019. Detection of Parkinson's disease through the peptoid recognizing α-synuclein in serum. ACS Chem. Neurosci. 10, 1204–1208.

Garcia Esteban, E., Cozar-Bernal, M.J., Rabasco Alvarez, A.M., Gonzalez-Rodriguez, M.L., 2018. A comparative study of stabilising effect and antioxidant activity of different antioxidants on levodopa-loaded liposomes. J. Microencapsul. 35, 357–371.

Georgieva, J.V., Hoekstra, D., Zuhorn, I.S., 2014. Smuggling drugs into the brain: an overview of ligands targeting transcytosis for drug delivery across the blood-brain barrier. Pharmaceutics 6, 557–583.

Gonzalez-Barrios, J., Bannon, M., Anaya-Martinez, V., Flores, G., 2006. Neurotensin polyplex as an efficient carrier for delivering the human GDNF gene into nigral dopamine neurons of hemiparkinsonian rats. Mol. Ther. 6, 857–865.

Gregoriadis, G., 2008. Liposome research in drug delivery: the early days. J. Drug Target. 16, 520–524.

Hans, M.L., Lowman, A.M., 2002. Biodegradable nanoparticles for drug delivery and targeting. Curr. Opin. Solid State Mater. Sci. 6, 319–327.

Haque, S., Md, S., Sahni, J.K., Ali, J., Baboota, S., 2014. Development and evaluation of brain targeted intranasal alginate nanoparticles for treatment of depression. J. Psychiatr. Res. 48, 1–12.

Hasadsri, L., Kreuter, J., Hattori, H., Iwasaki, T., George, J.M., 2009. Functional protein delivery into neurons using polymeric nanoparticles. J. Biol. Chem. 284, 6972–6981.

Hernandez-Chan, N., Bannon, M., Orozco-Barrios, C., Escobedo, L., Zamudio, S., Cruz, F.D.I., et al., 2015. Neurotensin-polyplex-mediated brain-derived neurotrophic factor gene delivery into nigral dopamine neurons prevents nigrostriatal degeneration in a rat model of early Parkinson's disease. J. Biomed. Sci. 22, 59.

Hebert, S.S., De Strooper, B., 2009. Alterations of the microRNA network cause neurodegenerative disease. Trends Neurosci. 32, 199–206.

Horst, C.H., Titze-de-Almeida, R., Titze-de-Almeida, S.S., 2017. The involvement of Eag1 potassium channels and miR-34a in rotenone-induced death of dopaminergic SH-SY5Y cells. Mol. Med. Rep. 15, 1479–1488.

Hollon, T., 2000. Researchers and regulators reflect on first gene therapy death. Nat. Med. 6, 6.

Huang, L., Liu, Y., 2011. In vivo delivery of RNAi with lipid-based nanoparticles. Annu. Rev. Biomed. Eng. 13, 507–530.

Husen, A., 2017. Gold nanoparticles from plant system: synthesis, characterization and application. In: Ghorbanpourn, M., Manika, K., Varma, A. (Eds.), Nanoscience and Plant–Soil SystemsVol. 48, 2017, vol. 11. Springer International Publishing AG, Gewerbestrasse, Cham, Switzerland, pp. 455–479, 6330.

Husen, A., Siddiqi, K.S., 2014. Phytosynthesis of nanoparticles: concept, controversy and application. Nano Res Lett 9, 229.

Hu, Q., Guo, F., Zhao, F., Fu, Z., 2017. Effects of titanium dioxide nanoparticles exposure on parkinsonism in zebrafish larvae and PC12. Chemosphere 173, 373–379.

Johnston, T.H., Fox, S.H., Brotchie, J.M., 2005. Advances in the delivery of treatments for Parkinson's disease. Expert Opin. Drug Deliv. 2, 1059–1073.

Juillerat-Jeanneret, L., 2008. The targeted delivery of cancer drugs across the blood-brain barrier: chemical modifications of drugs or drug-nanoparticles? Drug Discov. Today 13, 1099–1106.

Kakkar, A.K., Dahiya, N., 2015. Management of Parkinson's disease: current and future pharmacotherapy. Eur. J. Pharmacol. 750, 74–81.

Katzenschlager, R., Hughes, A., Evans, A., Manson, A.J., Hoffman, M., Swinn, L., et al., 2005. Continuous subcutaneous apomorphine therapy improves dyskinesias in Parkinson's disease: a prospective study using single-dose challenges. Mov. Disord. 20, 151–157.

Klyachko, N.L., Haney, M.J., Zhao, Y., Manickam, D.S., Mahajan, V., Suresh, P., et al., 2014. Macrophages offer a paradigm switch for CNS delivery of therapeutic proteins. Nanomedicine 9, 403–422.

Kumari, S., Ahsan, S.M., Kumar, J.M., Kondapi, A.K., Rao, N.M., 2017. Overcoming blood brain barrier with a dual purpose Temozolomide loaded Lactoferrin nanoparticles for combating glioma (SERP-17-12433). Sci. Rep. 7, 6602.

Kumar, P., Lakshmi, Y.S., C, B., Golla, K., Kondapi, A.K., 2015. Improved safety, bioavailability and pharmacokinetics of zidovudine through lactoferrin nanoparticles during oral administration in rats. PLoS One 10, e0140399.

Lalani, J., Baradia, D., Lalani, R., Misra, A., 2014. Brain targeted intranasal delivery of tramadol: comparative study of microemulsion and nanoemulsion. Pharm. Dev. Technol. 1–10.

Locatelli, E., Franchini, M.C., 2012. Biodegradable PLGA-b-PEG polymeric nanoparticles: synthesis, properties, and nanomedical applications as drug delivery system. J. Nanoparticle Res. 14, 1316.

Lara, H.H., Garza-Trevino, E.H., Ixtepan-Turrent, L., Singh, D.K., 2011. Silver nanoparticles are broad-spectrum bactericidal and virucidal compounds. J. Nanobiotechnol. 9, 30.

Lauzon, M.A., Daviau, A., Marcos, B., Faucheux, N., 2015. Nanoparticle mediated growth factor delivery systems: a new way to treat Alzheimer's disease. J. Control. Release 206, 187–205.

Malarkey, E.B., Parpura, V., 2007. Applications of carbon nanotubes in neurobiology. Neurodegener. Dis. 4, 292–299.

Maria, E., Aranda-Barradas, I.D., Marquez, M., Quintanar, L., Santoyo-Salazar, J., Armando, J., et al., 2018. Development of a parenteral formulation of NTS-polyplex nanoparticles for clinical purpose. Pharmaceutics 10, 5.

Martinez-Fong, D., Navarro-Quiroga, I., Ochoa, I., Alvarez-Maya, I., Meraz, M.A., Luna, J., et al., 2009. Neurotensin-SPDP-poly-L-lysine conjugate: a nonviral vector for targeted gene delivery to neural cells. Mol. Brain Res. 69, 249–262.

Masserini, M., 2013. Nanoparticles for brain drug delivery. ISRN Biochem 238428.

Mena, M.A., Garcia de Yebenes, J., 2008. Glial cells as players in parkinsonism: the "good," the "bad," and the "mysterious" glia. The Neuroscientist 14, 544–560.

Mingozzi, F., High, K.A., 2013. Immune responses to AAV vectors: overcoming barriers to successful gene therapy. Blood 122, 23–36.

Mittal, D., Md, S., Hasan, Q., Fazil, M., Ali, A., Baboota, S., et al., 2016. Brain targeted nanoparticulate drug delivery system of rasagiline via intranasal route. Drug Deliv. 23, 130–139.

Mohammadi, S., Nikkhah, M., 2017. TiO2 nanoparticles as potential promoting agents of fibrillation of α-synuclein, a Parkinson's disease-related protein. Iran. J. Biotechnol. 15, e1519.

Muntimadugu, E., Dhommati, R., Jain, A., Challa, V.G., Shaheen, M., Khan, W., 2016. Intranasal delivery of nanoparticle encapsulated tarenflurbil: a potential brain targeting strategy for Alzheimer's disease. Eur. J. Pharm. Sci. 92, 224–234.

Natarajan, G., Leibowitz, J.A., Zhou, J., Zhao, Y., McElroy, J.A., King, M.A., et al., 2017. Adeno-associated viral vector-mediated preprosomatostatin expression suppresses induced seizures in kindled rats. Epilepsy Res. 130, 81–92.

Neuwelt, E.A., Bauer, B., Fahlke, C., Fricker, G., Iadecola, C., Janigro, D., et al., 2011. Engaging neuroscience to advance translational research in brain barrier biology. Nat. Rev. Neurosci. 12, 169–182.

Nyholm, D., Nilsson Remahl, A.I., Dizdar, N., Constantinescu, R., Holmberg, B., Jansson, R., et al., 2005. Duodenal levodopa infusion monotherapy vs oral polypharmacy in advanced Parkinson disease. Neurology 64, 216–223, 2005.

Obulesu, M., Muralidhara Rao., Dowlathabad, Shamasundar, N.M., 2009. Studies on Genomic DNA stability in Aluminium Malotolate treated aged New Zealand rabbit: Relevance to the Alzheimer's Animal Model. J Clin. Med Res. 1, 212–218.

Obulesu, M., Jhansilakshmi, M., 2016. Neuroprotective role of Nanoparticles against Alzheimer's Disease. Curr Drug Metab 17, 142–149.

Obulesu, M., 2018. Multifarious therapeutic avenues for Alzheimer's disease. In: Singh, S., Joshi, N. (Eds.), In Pathology, Prevention and Therapeutics of Neurodegenerative Disorders. Springer publishers, p. 2018.

Obulesu, M., Muralidhara Rao, D., 2010. DNA damage and impairment of DNA repair in Alzheimer's Disease. Int. J. Neurosci. 120, 397–403.

Obeso, J.A., Rodriguez-Oroz, M.C., Goetz, C.G., Marin, C., Kordower, J.H., Rodriguez, M., et al., 2010. Missing pieces in the Parkinson's disease puzzle. Nat. Med. 16, 653–661.

Obulesu, M., Venu, R., Somashekhar, R., 2011a. Lipid peroxidation in Alzheimer's Disease: emphasis on metal mediated neurotoxicity. Acta Neurol. Scand. 124, 295–301.

Obulesu, M., Somashekhar, R., Venu, R., 2011b. Genetics of Alzheimer's disease: apo E and presenilins instigated neurodegeneration. Int. J. Neurosci. 121, 229–236.

Obulesu, M., Jhansilakshmi, M., 2014a. Apoptosis in Alzheimer's Disease: an understanding of the physiology, pathology and therapeutic avenues. Neurochem. Res. 39, 2301–2312.

Obulesu, M., Jhansilakshmi, M., 2014b. Neuroinflammation in Alzheimer's disease: an understanding of physiology and pathology. Int. J. Neurosci. 124, 227–235.

Orr, C.F., Rowe, D.B., Halliday, G.M., 2002. An inflammatory review of Parkinson's disease. Prog. Neurobiol. 68, 325–340.

Perez-Martinez, F.C., Carrion, B., Cena, V., 2012. The use of nanoparticles for gene therapy in the nervous system. J. Alzheimer's Dis. 31, 697–710.

Persidsky, Y., Ramirez, S.H., Haorah, J., Kanmogne, G.D., 2006. Blood-brain barrier: structural components and function under physiologic and pathologic conditions. J. Neuroimmune Pharmacol. 1, 223–236.

Prathipati, P., Zhu, J., Dong, X., 2016. Development of novel HDL-mimicking alpha-tocopherol-coated nanoparticles to encapsulate nerve growth factor and evaluation of biodistribution. Eur. J. Pharm. Biopharm. 108, 126–135.

Pringsheim, T., Jette, N., Frolkis, A., Steeves, T.D.L., 2014. The prevalence of Parkinson's disease: a systematic review and meta-analysis. Mov. Disord. 29, 1583–1590.

Razgado-Hernandez, L., Espadas-Alvarez, A., Reyna-Velazquez, P., Sierra-Sanchez, A., Anaya-Martinez, V., Jimenez-Estrada, I., et al., 2015. The transfection of BDNF to dopamine neurons potentiates the effect of dopamine D3 receptor agonist recovering the striatal innervation, dendritic spines and motor behavior in an aged rat model of Parkinson's disease. PLoS One 10, e0117391.

Ramanathan, S., Archunan, G., Sivakumar, M., Tamil Selvan, S., Fred, A.L., Kumar, S., 2018. Theranostic applications of nanoparticles in neurodegenerative disorders. Int J Nanomedicine 13, 5561–5576.

Song, Q., Song, H., Xu, J., Huang, J., Hu, M., Gu, X., et al., 2016. Biomimetic ApoE-reconstituted high density lipoprotein nanocarrier for blood-brain barrier penetration and amyloid beta-targeting drug delivery. Mol. Pharm. 13, 3976–3987.

Sikkandhar, M.G., Nedumaran, A.M., Ravichandar, R., Singh, S., Santhakumar, I., Goh, Z.C., et al., 2017. Theranostic probes for targeting tumor microenvironment: an overview. Int. J. Mol. Sci. 18, E1036.

Sridhar, S., Mishra, S., Gulyas, M., Padmanabhan, P., Gulyas, B., 2017. An overview of multimodal neuroimaging using nanoprobes. Int. J. Mol. Sci. 18, 311.

Sharma, V.K., Yngard, R.A., Lin, Y., 2009. Silver nanoparticles: green synthesis and their antimicrobial activities. Adv. Colloid Interface Sci. 145, 83–96.

Siddiqi, K.S., Husen, A., 2017. Recent advances in plant-mediated engineered gold nanoparticles and their application in biological system. J. Trace Elem. Med. Biol. 40, 10–23.

Siddiqi, K.S., Husen, A., Sohrab, S.S., Yassin, M.O., 2018. Recent Status of Nanomaterial Fabrication and Their Potential Applications in Neurological Disease Management. Nanoscale Res Lett 13, 231.

Siddiqi, K.S., Husen, A., 2016. Green synthesis, characterization and uses of palladium/platinum nanoparticles. Nano Res Lett 11, 482.

Spuch, C., Navarro, C., 2011. Liposomes for targeted delivery of active agents against neurodegenerative diseases (Alzheimer's disease and Parkinson's disease). J Drug Deliv 469679.

Stockwell, K.A., Scheller, D.K., Smith, L.A., Rose, S., Iravani, M.M., Jackson, M.J., et al., 2010. Continuous rotigotine administration reduces dyskinesia resulting from pulsatile treatment with rotigotine or L-DOPA in MPTP-treated common marmosets. Exp. Neurol. 221, 79–85.

Tanabe, S., Inoue, K.I., Tsuge, H., Uezono, S., Nagaya, K., Fujiwara, M., et al., 2017. The use of an optimized chimeric envelope glycoprotein enhances the efficiency of retrograde gene transfer of a pseudotyped lentiviral vector in the primate brain. Neurosci. Res. 120, 45–52.

Titze de Almeida, S.S., Horst, C.H., Soto-Sanchez, C., Fernandez, E., Titze de Almeida, R., 2018. Delivery of miRNA-targeted oligonucleotides in the rat striatum by magnetofection with neuromag. Molecules 23, 1825.

The Royal Society & The Royal Academy of Engineering, 2004. Nanomanufacturing and the industrial application of nanotechnologies. Available from: http://www.nanotec.org.uk/report/chapter4.pdf.

Vagner, T., Dvorzhak, A., Wojtowicz, A.M., Harms, C., Grantyn, R., 2016. Systemic application of AAV vectors targeting GFAP-expressing astrocytes in Z-Q175-KI Huntington's disease mice. Mol. Cell. Neurosci. 77, 76–86.

Vashist, A., Kaushik, A., Vashist, A., Bala, J., Nikkhah-Moshaie, R., Sagar, V., et al., 2018. Nanogels Potential Drug Nanocarriers for CNS Drug Delivery. Drug Discov. Today (in press).

Weksler, B.B., Subileau, E.A., Perriere, N., Charneau, P., Holloway, K., Leveque, M., et al., 2005. Blood-brain barrier-specific properties of a human adult brain endothelial cell line. FASEB J. 19, 1872–1874.

Wilson, B., 2009. Brain targeting PBCA nanoparticles and the blood-brain barrier. Nanomedicine 4, 499–502.

Xu, P., Gullotti, E., Tong, L., Highley, C.B., Errabelli, D.R., Hasan, T., et al., 2009. Intracellular drug delivery by poly(lactic-co-glycolic acid) nanoparticles, revisited. Mol. Pharm. 6, 190–201.

Yan, X., Xu, L., Bi, C., Duan, D., Chu, L., Yu, X., et al., 2018a. Lactoferrin-modified rotigotine nanoparticles for enhanced nose-to-brain delivery: LESA-MS/MS-based drug biodistribution, pharmacodynamics, and neuroprotective effects. Int. J. Nanomed. 13, 273–281.

Yan, A., Zhang, Y., Lin, J., Song, L., Wang, X., Liu, Z., 2018b. Partial depletion of peripheral M1 macrophages reverses motor deficits in MPTP-treated mouse by suppressing neuroinflammation and dopaminergic neurodegeneration. Front. Aging Neurosci. 10, 160, 2018.

Yu, X., Yao, J.Y., He, J., Tian, J.W., 2015. Protection of MPTP-induced neuroinflammation and neurodegeneration by rotigotine-loaded microspheres. Life Sci. 124, 136–143.

Zhang, H., Yang, B., Mu, X., Ahmed, S.S., Su, Q., He, R., et al., 2011. Several rAAV vectors efficiently cross the blood-brain barrier and transduce neurons and astrocytes in the neonatal mouse central nervous system. Mol. Ther. 19, 1440–1448.

Zhu, J., Dong, X., 2017. Preparation and characterization of novel hdl-mimicking nanoparticles for nerve growth factor encapsulation. J. Vis. Exp. 123, e55584.

Further reading

Joshi, P., Zhou, Y., Ahmadov, T.O., Zhang, P., 2014. Quantitative sers-based detection using Ag–Fe3O4 nanocomposites with an internal reference. J. Mater. Chem. C 2, 9964–9968.

CHAPTER 2

Biomaterials: a boon or a bane in the treatment of Parkinson's disease

1. Introduction

Neurodegenerative diseases with a few similar pathological events contribute for significant high death rate currently (Gendelman et al., 2015; Legname, 2015; Yacoubian, 2017; Kevadiya et al., 2018). Next to Alzheimer's disease (AD), Parkinson's disease (PD) is more prevalent with more than a million patients in the United States currently (Demaagd and Philip, 2015; Kevadiya et al., 2018). James Parkinson, an English surgeon for the first time identified PD in 1817, explained the same in his monograph entitled "An essay on the Shaking Palsy" [Pearce, 1989, Parkinson, 2002, Kevadiya et al., 2018].

Although PD most commonly affects 2% population over 65 years and 4% population over 85 years age, yet it is observed in young people also (McDonald et al., 2018; Sadowska-Bartosz and Bartosz, 2018). Bradykinesia, tremor of hands, legs, jaws, firmness of limbs and trunk, and perturbed balance and coordination are the symptoms of PD (Sadowska-Bartosz and Bartosz, 2018). PD etiology also involves accumulation of α-synuclein and parkin, resulting in the formation of Lewy bodies in cytoplasm (Luong and Nguyen, 2012; Sadowska-Bartosz and Bartosz, 2018). Moreover, metal dyshomeostasis has long been implicated in the etiology of neurodegenerative diseases such as AD and PD (Obulesu et al., 2011; Davies et al., 2016; Sadowska-Bartosz and Bartosz, 2018).

2. Oxidative stress

Protein nitration caused by reactions of peroxynitrite (ONOO-) with amino acids such as tyrosine, phenylalanine, and histidine is also considered

Parkinson's Disease Therapeutics
ISBN 978-0-12-819882-7
https://doi.org/10.1016/B978-0-12-819882-7.00002-7

as a vital event in the etiology of neurodegenerative diseases such as AD and PD (Aldini et al., 2013; Kim et al., 2015; Huang et al., 2016; Sadowska-Bartosz and Bartosz, 2018). Additionally, enhanced level of nitration associated with neuroinflammation has been observed in plethora of diseases such as AD, PD, amyotrophic lateral sclerosis, diabetes, rheumatoid arthritis, lupus, atherosclerosis, hypertension, and liver and cardiovascular diseases (Adams et al., 2015; Sadowska-Bartosz and Bartosz, 2018). To combat the oxidative stress, manifold dietary antioxidants were employed but they offered only a minimal protection (Ho et al., 2012; Ferreira et al., 2015; Sadowska-Bartosz and Bartosz, 2018). Despite the progress in neuroscience research for a few decades, there is no appropriate therapy till date (Dickson et al., 1999; Spillantini and Goedert, 2000; Galvin et al., 2001; Rockenstein et al., 2018).

3. Nanoparticles

The extensively used drug delivery systems (DDS) for brain targeting currently include polymeric nanoparticles (NPs), liposomes, inorganic NPs such as iron oxide NPs, and quantum dots (Sadowska-Bartosz and Bartosz, 2018). The vital components of NPs include outer surface layer that can be functionalized with multifarious small molecules, surfactants, metal ions, and polymers; the shell layer is made up of different chemicals from the core and the core which is the actual NP (Laurent et al., 2010; Shin et al., 2016; Sadowska-Bartosz and Bartosz, 2018). While a few delivery systems show promising efficacy, neurotoxicity of NPs on neurons and blood—brain barrier (BBB) are worth considering issues (Cupaioli et al., 2014; Sadowska-Bartosz and Bartosz, 2018).

Although BBB is a major constraint in the development of brain-targeted therapeutics, yet inorganic NPs showed promise in the treatment of neurodegenerative diseases (Busquets et al., 2014; Dilnawaz and Sahoo, 2015; Niu et al., 2017; Kevadiya et al., 2018). On similar lines, drug-loaded magnetic NPs were used to overcome neurodegenerative disorders such as AD and PD (Amiri et al., 2013; Ji et al., 2017; Kevadiya et al., 2018). Among multifarious nanomaterials in use, quantum dots, cerium oxide, yttrium oxide, molybdenum disulfide, and graphene-based materials have been found highly efficacious in diagnosis of AD and PD (Schubert et al., 2006; Estevez et al., 2011; Li et al., 2012; Das et al., 2013; Hirst et al., 2013; Zhao et al., 2016; Li et al., 2017; Naz et al., 2017; Vilela et al., 2017; Liu et al., 2018; Kevadiya et al., 2018).

Although L-DOPA is most commonly used in the treatment of PD, approximately 50% of patients using this treatment show serious complications (Katzenschlager and Lees, 2002; Stocchi, 2005; Kevadiya et al., 2018). To overcome these challenges, dopamine-loaded PLGA NPs were designed, which enhanced drug permeation through BBB after systemic intravenous injection in a 6-OHDA-instigated rat model of PD (Pahuja et al., 2015; Kevadiya et al., 2018). In another study, lactoferrin-tethered poly(ethylene glycol) PEG-PLGA NPs were designed, which spanned BBB via clathrin-mediated endocytosis and ameliorated the striatal lesions in rats (Kevadiya et al., 2018). In addition, pramipexole dihydrochloride—loaded chitosan NPs improved antioxidant ability of superoxide dismutase and catalase activities in turn improving the dopamine in PD rat brain (Raj et al., 2018; Kevadiya et al., 2018). Plethora of compounds commonly used to overcome intricate BBB include PEG, thiamine, transferrin, and glutathione (Gref et al., 2000; Oyewumi et al., 2003; Chang et al., 2009; Grover et al., 2014; Kevadiya et al., 2018).

4. Neurturin

Neurturin (NRTN) that shows vital effect on recovering dopaminergic system has been intranigrostriatally transfected into the brain using neurotensin (NTS)—polyplex nanoparticulate system [Reyes-Corona et al., 2017]. These NPs have been successfully used to accomplish NRTN gene delivery in dopaminergic neurons of the substantia nigra both in vitro [Hernandez-Baltazar et al., 2012, Reyes-Corona et al., 2017] and in vivo [Gonzalez-Barrios et al., 2006, Hernandez-Can et al., 2015, Razgado-Hernandez et al., 2015, Espadas-Alvarez et al., 2017, Reyes-Corona et al., 2017]. NTS—polyplex exhibits enhanced expression of NTS receptor type1 (NTSR1) in the plasma membrane of dopaminergic neurons, thus facilitating gene transfer by internalization of NTSR1 [Alvarez-Maya et al., 2001; Navarro-Quiroga et al., 2002; Martinez-Fong et al., 2012; Hernandez-Baltazar et al., 2012, Reyes-Corona et al., 2017].

5. Delivery systems for natural compounds

Growing body of evidence shows the substantial need for delivery systems to target natural compounds as they repose a few problems such as short systemic circulation and low bioavailability. In a recent study, Mead et al. (2017) designed brain permeating NPs, which successfully lead glial cell line—derived neurotrophic factor plasmid (~4 kb) and exhibited expression in target tissues in which MRI was employed with focused ultrasound (Kevadiya et al., 2018).

Mounting evidence has also shown that resveratrol, a potential plant-based polyphenolic compound, exerts neuroprotection by curtailing oxidative and ameliorating mitochondrial dysfunction in PD (Jin et al., 2008; Kevadiya et al., 2018). However, significantly less aqueous solubility and deliberate dissolution rate were the challenges associated with the use of resveratrol. To overcome these challenges, resveratrol NPs were designed, which profoundly improved the bioavailability, reduced oxidative stress, and improved mitochondrial dysfunction in rotenone-induced PD (Palle and Neerati, 2018, Kevadiya et al., 2018). On similar lines, Brenza et al. (2017) developed folic acid—functionalized mitochondria-targeted apocynin (Mito-apo) NPs, which remarkably attenuated H_2O_2 instigated cell death (Kevadiya et al., 2018).

Wealth of studies showed the interaction of epigallocatechin-3-gallate (EGCG), the most common polyphenolic compound of green tea with amyloid proteins such as α-synuclein (Weinreb et al., 2009; Bieschke et al., 2010; Xu et al., 2016; Kevadiya et al., 2018). EGCG is also an amenable DDS because it can reorganize amyloid formation pathways and provoke association of low-toxicity aggregates (Siddiqui and Mukhtar, 2010; Kevadiya et al., 2018). Nevertheless, it has low aqueous solubility, which impedes its bioavailability and cellular uptake (Kevadiya et al., 2018). To circumvent these challenges, selenium-tethered EGCG NPs were designed and conjugated to Tet-1 peptide, which offers profound affinity to neurons [Tet-1 EGCG@SeNPs]. These NPs showed considerable reduction in Aβ fibrillation and transformation of β-sheet fibril into amorphous aggregates in in vitro (Zhang et al., 2014; Kevadiya et al., 2018). Retinoic acid (RA) plays a key role in midbrain dopaminergic neurons due to the presence of its receptors and profound expression of RA synthesizing enzymes in neurons (Pan et al., 2019). In line with this, studies have shown the neuroprotective efficacy of RA-loaded NPs in in vitro and in vivo mouse model of PD (Esteves et al., 2015; Kevadiya et al., 2018).

6. Diagnosis

Manganese oxide NPs conjugated to L-DOPA showed sustained release of Mn^{2+} ions and L-DOPA. Because Mn^{2+} plays a pivotal role as a positive contrast in MRI, these NPs were found to facilitate both diagnosis and therapy of PD (McDonagh et al., 2016).

7. Silver nanoparticles

Although metallic materials such as gold and platinum are extensively used in electrochemical detection of biomolecules, low sensitivity at lower dopamine levels impeded their success (Yu et al., 2003; Zeis et al., 2007; Guo et al., 2010; Choo et al., 2017; Shin et al., 2017). To circumvent these challenges, Shin et al. (2017) for the first time designed a silver NP (SNP)—conjugated electrode enveloped by graphene oxide for sensing the dopamine. In this experiment, the surface of an indium tin oxide was altered with SNP graphene oxide by electrochemical deposition method. SNPs are inexpensive metallic materials and offer significantly high efficacy compared with gold and platinum in electrochemical detection. In addition, graphene oxide is available in plenty and exhibit better electrocatalytic ability (Shin et al., 2017). Therefore, they are amalgamated to design highly efficacious SNPs (Shin et al., 2017).

8. Immunotherapy

Vaccination with glucan microparticle (GP)+rapamycin (RAP)/α-synuclein has been found to generate enhanced levels of anti-α-synuclein antibodies. In this technique, humoral and immunosuppressive cellular immunizations were associated to accomplish α-synuclein scavenging. Therefore, to overcome α-synucleinopathies, a vaccine delivery system with antigen-presenting cell and targeting glucan microparticle codelivered with rapamycin exhibited 30%—45% decrease in α-synuclein aggregation and neuroinflammation (Rockenstein et al., 2018).

9. Shortcomings of nanotechnology

BBB-targeted NPs encounter following biological riddles: (1) protein and enzyme intervention with NPs, thus altering them and initiating their nonspecific localization (Aggarwal et al., 2009; Walczyk et al., 2010; Shah et al., 2012), (2) intricate NP fluid dynamics in blood vessels and vulnerability of target molecules on NPs (Blanco et al., 2015), (3) lower level and difference of cell-targeted biomarkers (Banks, 2009), extremely specific barriers (Duvernoy et al., 1983; Sarin, 2010), and a few specific transporters for specific cells, (4) variations in pH, tissues, and targeted organs (Sharma et al., 2018; Kevadiya et al., 2018). Despite remarkable protective efficacy of CeO_2 NPs, cerium released from these NPs has been found to exert toxic effects to normal cells and tissues in in vitro, which in turn leads to apoptosis (Park et al., 2008; Sadowska-Bartosz and Bartosz G, 2018).

10. Conclusions and future perspectives

Despite neuroprotective role of NPs, BBB poses a potential challenge with its tightly packed cell-to-cell contacts and blood capillary network (Shadab et al., 2009; Sweeney et al., 2018; Kevadiya et al., 2018). Moreover, insufficient techniques to detect neurodegenerative diseases at an early stage are also an important impediment in therapeutic intervention (Kevadiya et al., 2018).

It is indeed a herculean task to conclude if the brain-targeted nanomaterials are a boon or a bane in the treatment of neurodegenerative diseases (Khanna et al., 2015). Voluminous studies forged to overcome PD have emphasized that the multidimensional therapeutic strategies garnered from pathology, biochemistry, biomaterials, and molecular medicine expertise may give a viable solution. Manifold NPs used as substantial theranostic tools need to be validated for their accuracy and efficacy.

References

Adams, L., Franco, M.C., Estevez, A.G., 2015. Reactive nitrogen species in cellular signaling. Exp. Biol. Med. (Maywood) 240, 711–717.

Aggarwal, P., Hall, J.B., McLeland, C.B., Dobrovolskaia, M.A., McNeil, S.E., 2009. Nanoparticle interaction with plasma proteins as it relates to particle biodistribution, biocompatibility and therapeutic efficacy. Adv. Drug Deliv. Rev. 61, 428–437.

Aldini, G., Vistoli, G., Stefek, M., Chondrogianni, N., Grune, T., Sereikaite, J., et al., 2013. Molecular strategies to prevent, inhibit, and degrade advanced glycoxidation and advanced lipoxidation end products. Free Radic. Res. 47 (Suppl. 1), 93–137.

Alvarez-Maya, I., Navarro-Quiroga, I., Meraz-Rios, M.A., Aceves, J., Martinez-Fong, D., 2001. In vivo gene transfer to dopamine neurons of rat substantia nigra via the high-affinity neurotensin receptor. Mol. Med. 7, 186–192.

Amiri, H., Saeidi, K., Borhani, P., Manafirad, A., Ghavami, M., Zerbi, V., 2013. Alzheimer's disease: pathophysiology and applications of magnetic nanoparticles as MRI theranostic agents. ACS Chem. Neurosci. 4, 1417–1429.

Banks, W.A., 2009. Characteristics of compounds that cross the blood-brain barrier. BMC Neurol. 9, 1–5.

Bieschke, J., Russ, J., Friedrich, R.P., Ehrnhoefer, D.E., Wobst, H., Neugebauer, K., et al., 2010. EGCG remodels mature α-synuclein and amyloid-β fibrils and reduces cellular toxicity. Proc. Natl. Acad. Sci. 107, 7710–7715.

Blanco, E., Shen, H., Ferrari, M., 2015. Principles of nanoparticle design for overcoming biological barriers to drug delivery. Nat. Biotechnol. 33, 941–951.

Brenza, T.M., Ghaisas, S., Ramirez, J.E.V., Harischandra, D., Anantharam, V., Kalyanaraman, B., et al., 2017. Neuronal protection against oxidative insult by polyanhydride nanoparticle-based mitochondria-targeted antioxidant therapy. Nanomedicine 13, 809–820.

Busquets, M.A., Sabate, R., Estelrich, J., 2014. Potential applications of magnetic particles to detect and treat Alzheimer's disease. Nanoscale Res. Lett. 9, 538.

Chang, J., Jallouli, Y., Kroubi, M., Yuan, X., Feng, W., Kang, C., et al., 2009. Characterization of endocytosis of transferrin-coated PLGA nanoparticles by the blood–brain barrier. Int. J. Pharm. 379, 285–292.

Choo, S.S., Kang, E.S., Song, I., Lee, D., Choi, J.W., Kim, T.H., 2017. Electrochemical detection of dopamine using 3D porous graphene oxide/gold nanoparticle composites. Sensors 17, 861.

Cupaioli, F.A., Zucca, F.A., Boraschi, D., Zecca, L., 2014. Engineered nanoparticles. How brain friendly is this new guest? Prog. Neurobiol. 119–120, 20–38.

Das, S., Dowding, J.M., Klump, K.E., McGinnis, J.F., Self, W., Seal, S., 2013. Cerium oxide nanoparticles: applications and prospects in nanomedicine. Nanomedicine (Lond.) 8, 1483–1508.

Davies, K.M., Mercer, J.F., Chen, N., Double, K.L., 2016. Copper dyshomeostasis in Parkinson's disease: implications for pathogenesis and indications for novel therapeutics. Clin. Sci. 130, 565–574.

Demaagd, G., Philip, A., 2015. Parkinson's disease and its management: part 1: disease entity, risk factors, pathophysiology, clinical presentation, and diagnosis. Pharm. Therapeut. 40, 504–532.

Dickson, D.W., Liu, W., Hardy, J., Farrer, M., Mehta, N., Uitti, R., et al., 1999. Widespread alterations of alpha-synuclein in multiple system atrophy. Am. J. Pathol. 155, 1241–1251.

Dilnawaz, F., Sahoo, S.K., 2015. Therapeutic approaches of magnetic nanoparticles for the central nervous system. Drug Discov. Today 20, 1256–1264.

Duvernoy, H., Delon, S., Vannson, J.L., 1983. The vascularization of the human cerebellar cortex. Brain Res. Bull. 11, 419–480.

Espadas-Alvarez, A.J., Bannon, M.J., Orozco-Barrios, C.E., Escobedo-Sanchez, L., Ayala-Davila, J., Reyes-Corona, D., et al., 2017. Regulation of human GDNF gene expression in nigral dopaminergic neurons using a new doxycycline-regulated NTS-polyplex nanoparticle system. Nanomed. Nanotechnol. Biol. Med. 13, 1363–1375.

Esteves, M., Cristovao, A.C., Saraiva, T., Rocha, S.M., Baltazar, G., Ferreira, L., et al., 2015. Retinoic acid-loaded polymeric nanoparticles induce neuroprotection in a mouse model for Parkinson's disease. Front. Aging Neurosci. 7, 20.

Estevez, A.Y., Pritchard, S., Harper, K., Aston, J.W., Lynch, A., Lucky, J.J., et al., 2011. Neuroprotective mechanisms of cerium oxide nanoparticles in a mouse hippocampal brain slice model of ischemia. Free Radic. Biol. Med. 51, 1155–1163.

Ferreira, M.E., de Vasconcelos, A.S., da Costa Vilhena, T., da Silva, T.L., da Silva Barbosa, A., Gomes, A.R., et al., 2015. Oxidative stress in Alzheimer's disease: should we keep trying antioxidant therapies? Cell. Mol. Neurobiol. 35, 595–614.

Galvin, J.E., Lee, V.M., Trojanowski, J.Q., 2001. Synucleinopathies: clinical and pathological implications. Arch. Neurol. 58, 186–190.

Gendelman, H.E., Anantharam, V., Bronich, T., Ghaisas, S., Jin, H., Kanthasamy, A.G., et al., 2015. Mallapragada, nanoneuromedicines for degenerative, inflammatory, and infectious nervous system diseases. Nanomedicine 11, 751–767.

Gonzalez-Barrios, J.A., Lindahl, M., Bannon, M.J., Anaya-Martinez, V., Flores, G., Navarro-Quiroga, I., et al., 2006. Neurotensin polyplex as an efficient carrier for delivering the human GDNF gene into nigral dopamine neurons of hemiparkinsonian rats. Mol. Ther. 14, 857–865.

Gref, R., Luck, M., Quellec, P., Marchand, M., Dellacherie, E., Harnisch, S., et al., 2000. 'Stealth' corona-core nanoparticles surfacemodified by polyethylene glycol (PEG): influences of the corona (PEG chain length and surface density) and of the core composition on phagocytic uptake and plasma protein adsorption. Colloids Surf. B: Biointerfaces 18, 301–313.

Grover, A., Hirani, A., Pathak, Y., Sutariya, V., 2014. Brain-targeted delivery of docetaxel by glutathione-coated nanoparticles for brain cancer. AAPS PharmSciTech 15, 1562–1568.

Guo, S., Wen, D., Zhai, Y., Dong, S., Wang, E., 2010. Platinum nanoparticle ensemble-on-graphene hybrid nanosheet: one-pot, rapid synthesis and used as new electrode material for electrochemical sensing. ACS Nano 4, 3959–3968.

Hernandez-Baltazar, D., Martinez-Fong, D., Trudeau, L.E., 2012. Optimizing NTS-polyplex as a tool for gene transfer to cultured dopamine neurons. PLoS One 7, e51341.

Hernandez-Chan, N.G., Bannon, M.J., Orozco-Barrios, C.E., Escobedo, L., Zamudio, S., De la Cruz, F., et al., 2015. Neurotensin-polyplex-mediated brain-derived neurotrophic factor gene delivery into nigral dopamine neurons prevents nigrostriatal degeneration in a rat model of early Parkinson's disease. J. Biomed. Sci. 22, 59.

Hirst, S.M., Karakoti, A., Singh, S., Self, W., Tyler, R., Seal, S., et al., 2013. Bio-distribution and in vivo antioxidant effects of cerium oxide nanoparticles in mice. Environ. Toxicol. 28, 107–118.

Ho, Y., Poon, D.C., Chan, T.F., Chang, R.C., 2012. From small to big molecules: how do we prevent and delay the progression of age-related neurodegeneration? Curr. Pharmaceut. Des. 18, 15–26.

Huang, W.J., Zhang, X., Chen, W.W., 2016. Role of oxidative stress in Alzheimer's disease. Biomed. Rep. 4, 519–522.

Ji, B., Wang, M., Gao, D., Xing, S., Li, L., Liu, L., et al., 2017. Combining nanoscale magnetic nimodipine liposomes with magnetic resonance image for Parkinson's disease targeting therapy. Nanomedicine 12, 237–253.

Jin, F., Wu, Q., Lu, Y.F., Gong, Q.H., Shi, J.S., 2008. Neuroprotective effect of resveratrol on 6-OHDA-induced Parkinson's disease in rats. Eur. J. Pharmacol. 600, 78–82.

Katzenschlager, R., Lees, A.J., 2002. Treatment of Parkinson's disease: levodopa as the first choice. J. Neurol. 249, ii19–ii24.

Kevadiya, B.D., Ottemann, B.M., Thomas, M.B., Mukadam, I., Nigam, S., McMillan, J., et al., 2018. Neurotheranostics as personalized medicines. Adv. Drug Deliv. Rev. S0169–409X (18), 30261–30268.

Khanna, P., Ong, C., Bay, B.H., Baeg, G.H., 2015. Nanotoxicity: an interplay of oxidative stress, inflammation and cell death. Nanomaterials (Basel). 5, 1163–1180.

Kim, G.H., Kim, J.E., Rhie, S.J., Yoon, S., 2015. The role of oxidative stress in neurodegenerative diseases. Exp. Neurobiol. 24, 325–340.

Laurent, S., Forge, D., Port, M., Roch, A., Robic, C., Vander Elst, L., et al., 2010. Magnetic iron oxide nanoparticles: synthesis, stabilization, vectorization, physicochemical characterizations, and biological applications. Chem. Rev. 108, 2064–2110.

Legname, G., 2015. Novel Approaches to Diagnosis and Therapy in Neurodegenerative Diseases. Springer International Publishing, Cham, pp. 155–158.

Li, X., Du, X., 2017. Molybdenum disulfide nanosheets supported Au-Pd bimetallic nanoparticles for non-enzymatic electrochemical sensing of hydrogen peroxide and glucose. Sens. Actuators B Chem. 239, 536–543.

Li, M., Yang, X., Ren, J., Qu, K., Qu, X., 2012. Using graphene oxide high near-infrared absorbance for photothermal treatment of Alzheimer's disease. Adv. Mater. 24, 1722–1728.

Liu, Y., Xu, L.P., Wang, Q., Yang, B., Zhang, X., 2018. Synergistic inhibitory effect of GQDs–tramiprosate covalent binding on amyloid aggregation. ACS Chem. Neurosci. 9, 817–823.

Luong, K.V., Nguyen, L.T., 2012. Thiamine and Parkinson's disease. J. Neurol. Sci. 316, 1–8.

Martinez-Fong, D., Bannon, M.J., Trudeau, L.E., Gonzalez-Barrios, J.A., Arango-Rodriguez, M.L., Hernandez-Chan, N.G., et al., 2012. NTS-Polyplex: a potential nanocarrier for neurotrophic therapy of Parkinson's disease. Nanomed. Nanotechnol. Biol. Med. 8, 1052–1069.

McDonagh, B.H., Singh, G., Hak, S., Bandyopadhyay, S., Augestad, I.L., Peddis, D., et al., 2016. L-DOPA-CoatedManganese oxide nanoparticles as dual MRI contrast agents and drug-delivery vehicles. Small 12, 301–306.

McDonald, C., Gordon, G., Hand, A., Walker, R.W., Fisher, J.M., 2018. 200 years of Parkinson's disease: what have we learnt from James Parkinson? Age Ageing 47, 209–214.

Mead, B.P., Kim, N., Miller, G.W., Hodges, D., Mastorakos, P., Klibanov, A.L., et al., 2017. Novel focused ultrasound gene therapy approach noninvasively restores dopaminergic neuron function in a rat Parkinson's disease model. Nano Lett. 17, 3533–3542.

Navarro-Quiroga, I., Antonio Gonzalez-Barrios, J., Barron-Moreno, F., Gonzalez-Bernal, V., Martinez-Arguelles, D.B., Martinez-Fong, D., 2002. Improved neurotensin-vector-mediated gene transfer by the coupling of hemagglutinin HA2 fusogenic peptide and Vp1 SV40 nuclear localization signal. Brain Res. Mol. Brain Res. 105, 86–97.

Naz, S., Beach, J., Heckert, B., Tummala, T., Pashchenko, O., Banerjee, T., et al., 2017. Cerium oxide nanoparticles: a 'radical' approach to neurodegenerative disease treatment. Nanomedicine (Lond.) 12, 545–553.

Niu, S., Zhang, L.K., Zhang, L., Zhuang, S., Zhan, X., Chen, W.Y., et al., 2017. Guan, inhibition by multifunctional magnetic nanoparticles loaded with alpha-synuclein RNAi plasmid in a Parkinson's disease model. Theranostics 7, 344–356.

Obulesu, M., Venu, R., Somashekhar, R., 2011. Lipid peroxidation in Alzheimer's disease: emphasis on metal mediated neurotoxicity. Acta Neurol. Scand. 124, 295–301.

Oyewumi, M.O., Liu, S., Moscow, J.A., Mumper, R.J., 2003. Specific association of thiamine-coated gadolinium nanoparticles with human breast cancer cells expressing thiamine transporters. Bioconjug. Chem. 14, 404–411.

Pahuja, R., Seth, K., Shukla, A., Shukla, R.K., Bhatnagar, P., Chauhan, L.K.S., et al., 2015. Trans-blood brain barrier delivery of dopamine-loaded nanoparticles reverses functional deficits in parkinsonian rats. ACS Nano 9, 4850–4871.

Pan, J., Yu, J., Sun, L., Xie, C., Chang, L., Wu, J., et al., 2019. ALDH1A1 regulates postsynaptic μ-opioid receptor expression in dorsal striatal projection neurons and mitigates dyskinesia through transsynaptic retinoic acid signaling. Sci. Rep. 9, 3602.

Park, E.J., Choi, J., Park, Y.K., Park, K., 2008. Oxidative stress induced by cerium oxide nanoparticles in cultured BEAS-2B cells. Toxicology 245, 90–100.

Parkinson, J., 2002. An essay on the Shaking Palsy. J. Neuropsychiatry Clin. Neurosci. 14, 223–236.

Pearce, J.M., 1989. Aspects of the history of Parkinson's disease. J. Neurol. Neurosurg. Psychiatry 52, 6–10.

Raj, R., Wairkar, S., Sridhar, V., Gaud, R., 2018. Pramipexole dihydrochloride loaded chitosan nanoparticles for nose to brain delivery: development, characterization and in vivo anti-Parkinson activity. Int. J. Biol. Macromol. 109, 27–35.

Razgado-Hernandez, L.F., Espadas-Alvarez, A.J., Reyna-Velazquez, P., Sierra-Sanchez, A., Anaya-Martinez, V., Jimenez-Estrada, I., et al., 2015. The transfection of BDNF to dopamine neurons potentiates the effect of dopamine d3 receptor agonist recovering the striatal innervation, dendritic spines and motor behavior in an aged rat model of Parkinson's disease. PLoS One 10, e0117391.

Reyes-Corona, D., Vazquez-Hernandez, N., Escobedo, L., Orozco-Barrios, C.E., Ayala-Davila, J., Moreno, M.G., et al., 2017. Neurturin overexpression in dopaminergic neurons induces presynaptic and postsynapticstructural changes in rats with chronic 6-hydroxydopamine lesion. PLoS One 27 (12), e0188239.

Rockenstein, E., Ostroff, G., Dikengil, F., Rus, F., Mante, M., Florio, J., et al., 2018. Combined active humoral and cellular immunization approaches for the treatment of synucleinopathies. J. Neurosci. 38, 1000–1014.

Sadowska-Bartosz, I., Bartosz, G., 2018. Redox nanoparticles: synthesis, properties and perspectives of use for treatment of neurodegenerative diseases. J. Nanobiotechnol. 16, 87.

Sarin, H., 2010. Physiologic upper limits of pore size of different blood capillary types and another perspective on the dual pore theory of microvascular permeability. J. Angiogenes Res. 2, 1–19.

Schubert, D., Dargusch, R., Raitano, J., Chan, S.W., 2006. Cerium and yttrium oxide nanoparticles are neuroprotective. Biochem. Biophys. Res. Commun. 342, 86–91.

Shadab, A.P., Zeenat, I., Syed, M.A.Z., Sushma, T., Divya, V., Gaurav, K.J., et al., 2009. CNS drug delivery systems: novel approaches. Recent Pat. Drug Deliv. Formul. 3, 71–89.

Shah, N.B., Vercellotti, G.M., White, J.G., Fegan, A., Wagner, C.R., Bischof, J.C., 2012. Blood–Nanoparticle interactions and in vivo biodistribution: impact of surface PEG and ligand properties. Mol. Pharm. 9, 2146–2155.

Sharma, A., Cornejo, C., Mihalic, J., Geyh, A., Bordelon, D.E., Korangath, P., et al., 2018. Physical characterization and in vivo organ distribution of coated iron oxide nanoparticles. Sci. Rep. 8, 4916.

Shin, W.K., Cho, J., Kannan, A.G., Lee, Y.S., Kim, D.W., 2016. Cross-linked composite gel polymer electrolyte using mesoporous methacrylate-functionalized SiO_2 nanoparticles for lithium-ion polymer batteries. Sci. Rep. 6, 26332.

Shin, J.W., Kim, K.J., Yoon, J., Jo, J., El-Said, W.A., Choi, J.W., 2017. Silver nanoparticle modified electrode covered by graphene oxide for the enhanced electrochemical detection of dopamine. Sensors (Basel) 17, E2771.

Siddiqui, I.A., Mukhtar, H., 2010. Nanochemoprevention by bioactive food components: a perspective. Pharm. Res. 27, 1054–1060.

Spillantini, M.G., Goedert, M., 2000. The alpha-synucleinopathies: Parkinson's disease, dementia with Lewy bodies, and multiple system atrophy. Ann. N.Y. Acad. Sci. 920, 16–27.

Stocchi, F., 2005. Optimising levodopa therapy for the management of Parkinson's disease. J. Neurol. 252, Iv43–iv48.

Sweeney, M.D., Sagare, A.P., Zlokovic, B.V., 2018. Blood–brain barrier breakdown in Alzheimer disease and other neurodegenerative disorders. Nat. Rev. Neurol. 14, 133.

Vilela, P., El-Sagheer, A., Millar, T.M., Brown, T., Muskens, O.L., Kanaras, A.G., 2017. Graphene oxide-upconversion nanoparticle based optical sensors for targeted detection of mRNA biomarkers present in Alzheimer's disease and prostate cancer. ACS Sens. 2, 52–56.

Walczyk, D., Bombelli, F.B., Monopoli, M.P., Lynch, I., Dawson, K.A., 2010. What the cell "Sees" in bionanoscience. J. Am. Chem. Soc. 132, 5761–5768.

Weinreb, O., Amit, T., Mandel, S., Youdim, M.B.H., 2009. Neuroprotective molecular mechanisms of (−)-epigallocatechin-3-gallate: a reflective outcome of its antioxidant, iron chelating and neuritogenic properties. Genes Nutr. 4, 283–296.

Xu, Y., Zhang, Y., Quan, Z., Wong, W., Guo, J., Zhang, R., et al., 2016. McGeer, H. Qing, epigallocatechin gallate (EGCG) inhibits alpha-synuclein aggregation: a potential agent for Parkinson's disease. Neurochem. Res. 41, 2788–2796.

Yacoubian, T.A., 2017. Chapter 1 – Neurodegenerative Disorders: Why Do We Need New Therapies? A2 – Adejare, Adeboye, Drug Discovery Approaches for the Treatment of Neurodegenerative Disorders. Academic Press, pp. 1–16.

Yu, A., Liang, Z., Cho, J., Caruso, F., 2003. Nanostructured electrochemical sensor based on dense gold nanoparticle films. Nano Lett. 3, 1203–1207.

Zeis, R., Mathur, A., Fritz, G., Lee, J., Erlebacher, J., 2007. Platinum-plated nanoporous gold: an efficient, low Pt loading electrocatalyst for PEM fuel cells. J. Power Sources 165, 65–72.

Zhang, J., Zhou, X., Yu, Q., Yang, L., Sun, D., Zhou, Y., et al., 2014. Epigallocatechin-3-gallate (EGCG)-Stabilized selenium nanoparticles coated with tet-1 peptide to reduce amyloid-β aggregation and cytotoxicity. ACS Appl. Mater. Interfaces 6, 8475–8487.

Zhao, Y., Xu, Q., Xu, W., Wang, D., Tan, J., Zhu, C., et al., 2016. Probing the molecular mechanism of cerium oxide nanoparticles in protecting against the neuronal cytotoxicity of a[small beta]1-42 with copper ions. Metallomics 8, 644–647.

Further reading

Clark, A.J., Davis, M.E., 2015. Increased brain uptake of targeted nanoparticles by adding an acid-cleavable linkage between transferrin and the nanoparticle core. Proc. Natl. Acad. Sci. 112, 12486–12491.

Hu, K., Shi, Y., Jiang, W., Han, J., Huang, S., Jiang, X., 2011. Lactoferrin conjugated PEG-PLGA nanoparticles for brain delivery: preparation, characterization and efficacy in Parkinson's disease. Int. J. Pharm. 415, 273–283.

Kwon, H.J., Cha, M.Y., Kim, D., Kim, D.K., Soh, M., Shin, et al., 2016. Mitochondria-targeting ceria nanoparticles as antioxidants for Alzheimer's disease. ACS Nano 10, 2860–2870.

CHAPTER 3

Blood—brain barrier targeted nanoparticles

1. Introduction

1.1 Physiology

Blood—brain barrier (BBB) plays a vital role in normal physiology and brain protection. Brain comprises around 644 km of blood vessels that deliver oxygen and energy compounds and eliminate carbon dioxide and metabolic wastes from the brain (Zlokovic, 2008; Kisler et al., 2017; Sweeney et al., 2018). Intact BBB mediates the chemical components of brain interstitial fluid, which in turn plays an essential role in information handling, synaptic functioning, and neuronal integrity (Zlokovic, 2011; Iadecola, 2004, 2013; Kisler et al., 2017; Sweeney et al., 2016; Sweeney et al., 2018). Commonly, in normal physiological conditions, molecular transport takes place through different mechanisms such as passive diffusion, carrier-mediated transport, adsorptive endocytosis, and receptor-mediated transport (Grabrucker et al., 2016, Fig. 3.1). BBB dysfunction augments vascular porosity, decreases cerebral blood flow, and disintegrates hemodynamic responses (Sweeney et al., 2018). Its impairment has long been implicated in neurodegenerative diseases such as Alzheimer's disease (AD) and Parkinson's disease (PD) (Miyakawa, 2010; Zlokovic, 2011; Baloyannis, 2015; Erdo et al., 2017; Obulays, 2018). The capillary walls of BBB isolate the brain from blood (Begley and Brightman, 2003; Olivier, 2005; Ramanathan et al., 2018). BBB restricts the passage of solutes into the brain as it is devoid of fenestrations or intracellular cleft (Ramanathan et al., 2018). Additionally, P-glycoprotein, multispecific organic anion transporter, and multidrug resistance proteins act as efflux pumps that drive away drugs despite their lipophilicity (Hans and Lowman, 2002; Persidsky et al., 2006; Ramanathan et al., 2018). Wealth of studies has shown that 98% of small molecules and approximately all large molecules fail to ferry BBB (Huang et al., 2016; Enteshari Najafabadi et al., 2018). Accordingly, BBB

Parkinson's Disease Therapeutics
ISBN 978-0-12-819882-7
https://doi.org/10.1016/B978-0-12-819882-7.00003-9

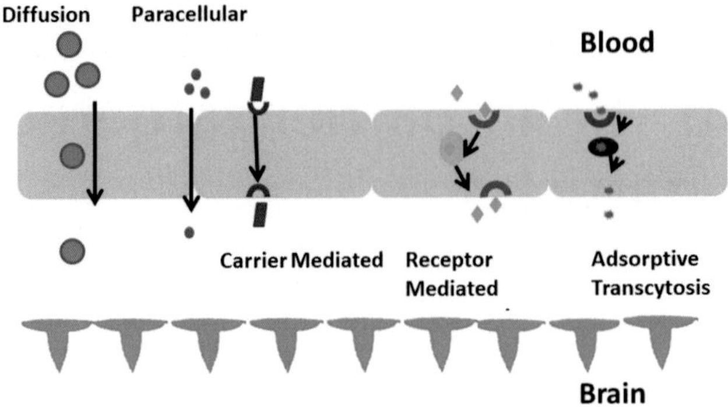

Figure 3.1 Multifarious pathways of nanoparticle entry through bllood brain barrier.

ferrying has long been a herculean task for therapeutic molecules (Neuwelt et al., 2008; Enteshari Najafabadi et al., 2018).

2. Blood–brain barrier impairment

BBB impairment results in the entry of neurotoxic blood-borne micro-organisms, unwanted components, and cells into the brain, which eventually instigate inflammation (Sweeney et al., 2018). In PD patients, BBB impairment has been understood by the perivascular accumulation of fibrinogen and fibrin (Gray and Woulfe, 2015), immunoglobulin G (IgG) (Pienaar et al., 2015), and hemosiderin in the striatal region (Loeffler et al., 1995; Gray and Woulfe, 2015; Sweeney et al., 2018). In line with this, endothelial disintegration with microvascular alteration and tight junction protein dysfunction is also observed in PD (Pienaar et al., 2015; Sweeney et al., 2018). Atypical angiogenesis as expressed by angiogenesis marker modification has been observed in substantia nigra, locus coeruleus, and putamen in PD (Wada et al., 2006; Desai Bradaric et al., 2012; Sweeney et al., 2018). Additionally, cerebrospinal fluid (CSF) biomarkers such as albumin quotient (Pisani et al., 2012; Janelidze et al., 2017) and CSF:serum IgG ratio were found to be enhanced in early-stage PD patients without dementia (Sweeney et al., 2018).

2.1 Nanoparticles

A drug needs to overcome several barriers before entering the brain such as BBB, blood CSF barrier, and efflux protein (Khan et al., 2018). Unfortunately, more than 90% of the brain-targeted novel drugs were not approved by the FDA (Cummings et al., 2015; Kang et al., 2018). Among multifarious techniques used till date to accomplish therapeutic delivery through BBB, nanotechnological strategies present significant promise (Khan et al., 2018; Obulesu, 2019). Nanotechnology has long been in use to facilitate the entry of theranostic molecules across BBB (Ramanathan et al., 2018). The advent of nanotechnology has brought diagnosis and therapeutics closer almost in a single agent (Ramanathan et al., 2018). Numerous nanoparticles (NPs) used to overcome BBB include liposomes, polymeric materials, and inorganic NPs, which show sustained drug release for a considerably longer duration (Begley and Kreuter, 1999; Juillerat-Jeanneret, 2008; Ramanathan et al., 2018). In addition, polymeric materials, liposomes, solid lipid NPs, and micelles have been extensively used in the treatment of neurodegenerative diseases because they exhibit significantly high therapeutic efficacy (Khan et al., 2018). The primary requirements for the NPs to accomplish effective BBB targeting are size less than 100 nm, biocompatibility, biodegradability, overcome reticuloendothelial system, and follow receptor-mediated transport (Ramanathan et al., 2018).

2.2 Cerium oxide nanoparticles

Although antioxidants have been of considerable importance in the t-reatment of neurodegenerative diseases such as AD and PD, yet their inability to ferry BBB impede their success. To overcome BBB, currently cerium oxide NPs (CeO_2) have been extensively used, which preferably eliminate reactive oxygen species (Nazem and Mansoori, 2011; Gandhi and Abramov, 2012; Hsieh and Yang, 2013; Dowding et al., 2013; Gupta et al., 2014; Behera et al., 2014; Naz et al., 2017). This is due to the cerium's switch between +3 and +4 oxidation states, which is similar to the redox activity of superoxide dismutase and catalase (Chen et al., 2006; Heckman et al., 2013; Estevez and Erlichman, 2014; Kumar et al., 2014; Pulido-Reyes et al., 2015; Walkey et al., 2015; Dahle and Arai, 2015; Naz et al., 2017).

To test the neuroprotective efficacy of CeONPs in PD, Pinna et al. (2015) investigated the antioxidative effect of size-regulated CeONPs in manganese-treated catecholaminergic cells (PC12) (Naz et al., 2017). Additionally, they also studied the combination of therapy of CeONPs and L-DOPA, which showed profound therapeutic efficacy compared to CeONPs alone (Naz et al., 2017). In another PD study of 1-methyl-4-phenyl-1,2,3,6-tetrahydropyridine(MPTP) mouse model, Dillon et al. (2011) reported that CeONPs substantially restored striatal dopamine levels and guarded dopaminergic neurons in the substantia nigra even at the lowest concentration of 0.5–5 μL (Naz et al., 2017). Despite the appreciable capability of spanning BBB and inducing therapeutic efficacy, CeONPs repose systemic toxicity by internalization in the liver, kidney, and spleen (Dan et al., 2012; Rojas et al., 2012; Yokel et al., 2013; Portioli et al., 2013; Graham et al., 2014; Naz et al., 2017). Therefore, there is a growing need to substantiate the therapeutic efficacy of CeONPs.

3. Blood–brain barrier spanning of nanoparticles

NPs ferry BBB in two pathways. The first pathway involves the central nervous system (CNS) localization of NPs via sensory nerve endings existing at airway epithelia or through the nerve endings of the olfactory bulb. Nevertheless, these findings are to be validated in human models and the safety concerns are to be addressed (Li et al., 2015; Ramanathan et al., 2018). The second pathway involves the drug introduction through nasal route, which successfully averts drug degradation, first-pass metabolism in the liver, and enhances bioavailability (Muller et al., 2001; Wilson et al., 2008; Kulkarni et al., 2010; Wilson et al., 2010; Ramanathan et al., 2018).

A few brain spanning natural polymeric NPs include chitosan, gelatin, and sodium alginate NPs, and synthetic polymers are poly(D,L-lactide-co-glycolide) (PLGA), poly(D,L-lactide), polybutyl cyanoacrylates, polycaprolactone, and poly(ethylenimine) NPs (Claudio et al., 2016; Khan et al., 2018). PLGA, which has FDA approval, has been an amenable biopolymer because of its safety profile, targeting ability, and ease of surface functionalization. Therefore, they are feasibly surface functionalized with surfactants or receptor targeting proteins, antibodies to accomplish brain-targeted delivery (Crawford et al., 2016; Khan et al., 2018). Albumin, which is adequately found in blood, has been extensively used in BBB targeting because of its amenable characteristics such as long half-life in blood, biocompatibility, and ability to trespass BBB effectively (Lee and Youn, 2016; Kang et al., 2018).

4. Parkinson's disease

Neurodegenerative disorders are the fourth deleterious disorders among the CNS diseases (Naz et al., 2017). PD is the most common neurodegenerative disorder in the elderly. Drugs such as apomorphine and rotigotine have been administered using subcutaneous method and patch delivery method to achieve uniform distribution in the epidermis (Bennett and Piercey, 1999; Johnston et al., 2005; Katzenschlager et al., 2005; Nyholm et al., 2005; Ramanathan et al., 2018). PD diagnosis has shown the low levels of astrocytes in substantia nigra, which is the result of downstream processing in the brain (Orr et al., 2002; Mena and Garcia de Yebenes, 2008; Obeso et al., 2010; Ramanathan et al., 2018).

5. Nanoparticles for Parkinson's disease

Yurek et al. (2009) designed NPs which specifically localized glial-derived neurotrophic factor–expressing gene in embryonic dopamine neurons, thus facilitating its overexpression and neurotrophic support (Ramanathan et al., 2018). On similar lines, lactoferrin NPs, which were synthesized by thiolating and tethering lactoferrin to pegylated NPs, enhanced odorranalectin modification and improved the therapeutic efficacy (Hu et al., 2011; Ramanathan et al., 2018). Levodopa (L-DOPA) spans BBB successfully using large neutral aminoacid transporter (LAT)-1 protein. Based on this, L-DOPA conjugated multibranched flower like gold NPs (L-DOPA-AuNFs) were synthesized by using catechols as reducing-cum-capping agent, which enhanced brain entry (Gonzalez-Carter et al., 2018).

6. Gold nanoparticles

Mounting evidence suggests that gold NPs (AuNPs) offer a few worth considering characteristics such as feasible functionalization, imaging, and increased biocompatibility, thus facilitating brain delivery of loaded therapeutics (Popovtzer et al., 2008; Alkilany and Murphy, 2010; Gonzalez-Carter et al., 2018). In line with this, AuNPs functionalized with transferrin (Wiley et al., 2013; Clark and Davis, 2015), insulin (Shilo et al., 2014), and glucose (Gromnicova et al., 2013) have been discovered (Gonzalez-Carter et al., 2018). Although these AuNPs are preferably internalized by microglial cells, microglial cell activation is not provoked, thus confirming the safety of NPs (Goode et al., 2015; Gonzalez-Carter et al., 2018). Despite their enhanced brain spanning, internalization in other organs leading to systemic toxicity limited their success (Gonzalez-Carter et al., 2018). In another study, AuNPs were altered with chitosan and tethered to plasmid DNA and nerve growth factor, which successfully delivered siRNA and downregulated genes encoding α-synuclein (Hu et al., 2018; Khan et al., 2018). Therefore, these NPs showed promise in the treatment of PD.

Glucose transport proteins on BBB endothelial cells have been profoundly used in internalization of NPs such as AuNPs, which showed remarkably high BBB transport (Gromnicova et al., 2013; Jiang et al., 2014; Patching, 2017; Kang et al., 2018). In another study, AuNPs conjugated with bicyclo(paraquat-p-phenylene) receptors to analyze dopamine and L-DOPA with a detection limit of 1×10^{-6} M (Baron et al., 2005; Kang et al., 2018). Aptamer-conjugated AuNPs tethered to electrode ameliorate the sensitivity upto 1×10^{-8} M, which facilitate the reduced dopamine detection in urine samples also (Xu et al., 2015; Kang et al., 2018).

7. Quercetin nanoparticles

Quercetin (3,3′,4′,5,7-pentahydroxyflavone), a potential bioactive flavonoid, shows profound neuroprotective effect against manifold neurodegenerative diseases (Enteshari Najafabadi et al., 2018). Additionally, it shows antioxidant (Ramos et al., 2006), antiinflammatory (Stewart et al., 2008), anticancer (Chakraborty et al., 2012), antiviral (Choi et al., 2009), and antiischemic effect (Duarte et al., 1993; Davis et al., 2009; Enteshari Najafabadi et al., 2018). Nevertheless, its low efficacy of BBB spanning limits its protective ability in the body. Therefore, superparamagnetic iron oxide nanoparticles (SPIONs) were designed, which showed incremental bioavailability about 10-folds (Enteshari Najafabadi et al., 2018). As these NPs show magnetic properties, they are extensively used in drug delivery, bioseparation, cell tracking, magnetic resonance imaging, and magnetic hyperthermia (Barreto et al., 2011; Chowdhury et al., 2017; Enteshari Najafabadi et al., 2018). Iron ions may perturb biomacromolecules via oxidative damage. Therefore, they are to be cautiously employed to overcome neurodegenerative diseases (Valdiglesias et al., 2015; Vinzant et al., 2017, Enteshari Najafabadi et al., 2018). Interestingly, these NPs show better sensitivity to brain cells compared to other organs' cells (Mahmoudi et al., 2011; Laurent et al., 2012; Enteshari Najafabadi et al., 2018). Although study has produced worth considering results, they are yet to be corroborated in a longer duration study (Enteshari Najafabadi et al., 2018). Table 3.1.

Table 3.1 List of recently studied Blood—brain barrier (BBB)—targeted Nanoparticles (NPs).

Model	Drug delivery system	Target	Result	Reference
Brain	CeO_2 NPs	Reactive oxygen species (ROS) scavenging	ROS scavenged	Nazem and Mansoori (2011), Gandhi and Abramov, 2012, Hsieh and Yang (2013), Dowding et al. (2013), Gupta et al. (2014), Behera et al. (2014), Naz et al. (2017)
PC12 cells	CeONPs	Antioxidative effect	Showed antioxidant effect	Pinna et al. (2015)
	CeONPs+L-DOPA	Manganese chloride induced oxidative stress reduction	Showed significantly enhanced protective efficacy compared to CeONPs alone	Naz et al. (2017)
MPTP mouse model	CeONPs	Restoration of dopamine levels	Dopamine levels restored	Dillon et al. (2011)
Embryonic dopamine neurons	Glial-derived neurotrophic factor (GDNF)—loaded NPs	Localization of GDNF in neurons	Enhanced expression of GDNF	Yurek et al. (2009)
Brain macrophages	L-DOPA-AuNFs	BBB	Enhanced entry into the brain	Gonzalez-Carter et al., 2018
Microglial cells	Gold NPs (AuNPs) functionalized with transferrin, insulin and glucose	BBB	Enhanced entry into the brain	Wiley et al. (2013), Clark and Davis (2015), Gromnicova et al., 2013; Goode et al. (2015), Gonzalez-Carter et al., 2018

Continued

Table 3.1 List of recently studied Blood—brain barrier (BBB)—targeted Nanoparticles (NPs).—cont'd

Model	Drug delivery system	Target	Result	Reference
PC12 cells	AuNPs tethered to plasmid DNA and nerve growth factor	siRNA delivery and downregulate α-synuclein genes	siRNA delivered and α-synuclein genes downregulated	Hu et al. (2018), Khan et al. (2018)
Dopaminergic neurons	AuNPs conjugated with bicyclo(paraquat-p-phenylene) receptors	Parkinson's disease diagnosis	Dopamine level detection	Baron et al. (2005), Xu et al. (2015), Kang et al. (2018)
Brain cells	Superparamagnetic iron oxide nanoparticles	Enhance quercetin (QT) bioavailability	QT bioavailability improved	Mahmoudi et al. (2011); Laurent et al. (2012); Enteshari Najafabadi et al., 2018

8. Conclusions and future perspectives

Although multifarious NPs show promise for the treatment of neurodegenerative diseases such as AD and PD, yet the underlying mechanisms of their metabolism are elusive (Enteshari Najafabadi et al., 2018). In addition, panoply of brain targets and hitherto unraveled targeted mechanisms yielded limited success (Ramanathan et al., 2018). Multifarious brain-targeted NPs showed localization in other organs and toxicity despite their significant brain penetration (Gonzalez-Carter et al., 2018). Several researchers are of the opinion that the impaired BBB during pathological conditions facilitate the entry of therapeutics feasibly. Nevertheless, impaired BBB with structurally and functionally altered blood vessels, endothelial disintegration, and decreased expression of tight junctions and adherens junctions remarkably hamper the therapeutic delivery to the brain (Sweeney et al., 2018). The molecular underpinnings involved in BBB impairment are elusive despite the voluminous studies aimed to unravel the same (Sweeney et al., 2018).

References

Alkilany, A., Murphy, C., 2010. Toxicity and cellular uptake of gold nanoparticles: what we have learned so far? J. Nanopart. Res. 12, 2313—2333.

Baloyannis, S.J., 2015. Brain capillaries in Alzheimer's disease. Hellenic J. Nucl. Med. 18, 152.

Barreto, A., Santiago, V., Mazzetto, S., Denardin, J.C., Lavin, R., Mele, G., et al., 2011. Magnetic nanoparticles for a new drug delivery system to control quercetin releasing for cancer chemotherapy. J. Nanopart. Res. 13, 6545—6553.

Baron, R., Zayats, M., Willner, I., 2005. Dopamine-, l-DOPA-, adrenaline-, and noradrenaline-induced growth of Au nanoparticles: assays for the detection of neurotransmitters and of tyrosinase activity. Anal. Chem. 77, 1566—1571.

Begley, D.J., Kreuter, J., 1999. Do ultra-low frequency (ULF) magnetic fields affect the blood-brain barrier? In: Holick, M.F., Jung, E.G. (Eds.), Biologic Effects of Light. Springer US, Boston, MA, pp. 297—301.

Begley, D.J., Brightman, M.W., 2003. Structural and functional aspects of the blood-brain barrier. Prog. Drug Res. 61, 39—78.

Behera, R., Goel, S., Das, S., Bouzahzah, B., Domann, N., Mahapatra, A., et al., 2014. Oxidative stress in Alzheimer's disease: targeting with nanotechnology. Biochem. Biophys. J. Neutron Ther. Cancer Treat. 2, 18—26.

Bennett, J.P., Piercey, M.F., 1999. Pramipexole — a new dopamine agonist for the treatment of Parkinson's disease. J. Neurol. Sci. 163, 25—31.

Chakraborty, S., Stalin, S., Das, N., Choudhury, S.T., Ghosh, S., Swarnakar, S., 2012. The use of nano-quercetin to arrest mitochondrial damage and MMP-9 upregulation during prevention of gastric inflammation induced by ethanol in rat. Biomaterials 33, 2991—3001.

Chen, J., Patil, S., Seal, S., McGinnis, J.F., 2006. Rare earth nanoparticles prevent retinal degeneration induced by intracellular peroxides. Nat. Nanotechnol. 1, 142—150.

Choi, H.J., Kim, J.H., Lee, C.H., Ahn, Y.J., Song, J.H., Baek, S.H., et al., 2009. Antiviral activity of quercetin 7-rhamnoside against porcine epidemic diarrhea virus. Antivir. Res. 81, 77–81.

Chowdhury, P., Nagesh, P.K., Kumar, S., Jaggi, M., Chauhan, S.C., Yallapu, M.M., 2017. Pluronic nanotechnology for overcoming drug resistance. In: Bioactivity of Engineered Nanoparticles. Springer, pp. 207–237.

Clark, A.J., Davis, M.E., 2015. Increased brain uptake of targeted nanoparticles by adding an acid-cleavable linkage between transferrin and the nanoparticle core. Proc. Natl. Acad. Sci. U.S.A. 112, 12486–12491.

Claudio, P., Reatul, K., Brigitte, E., Geraldine, P., 2016. Drug-delivery nanocarriers to cross the blood–brain barrier. Nanobiomater. Drug Deliv. 9, 333–370.

Crawford, L., Rosch, J., Putnam, D., 2016. Concepts, technologies, and practices for drug delivery past the blood-brain barrier to the central nervous system. J. Control. Release 240, 251–266.

Cummings, J., Mintzer, J., Brodaty, H., Sano, M., Banerjee, S., Devanand, D.P., et al., 2015. Agitation in cognitive disorders: International Psychogeriatric Association provisional consensus clinical and research definition. Int. Psychogeriatr. 27, 7–17.

Dahle, J.T., Arai, Y., 2015. Environmental geochemistry of cerium: applications and toxicology of cerium oxide nanoparticles. Int. J. Environ. Res. Public Health 12, 1253–1278.

Dan, M., Tseng, M.T., Wu, P., Unrine, J.M., Grulke, E.A., Yokel, A., 2012. Brain microvascular endothelial cell association and distribution of a 5 nm ceria engineered nanomaterial. Int. J. Nanomed. 7, 4023–4036.

Davis, J.M., Murphy, E.A., Carmichael, M.D., Davis, B., 2009. Quercetin increases brain and muscle mitochondrial biogenesis and exercise tolerance. Am. J. Physiol. Regul. Integr. Comp. Physiol. 296, R1071–R1077.

Dowding, J.M., Seal, S., Self, W.T., 2013. Cerium oxide nanoparticles accelerate the decay of peroxynitrite (ONOO(-)). Drug Deliv. Transl. Res. 3, 375–379.

Desai Bradaric, B., Patel, A., Schneider, J.A., Carvey, P.M., Hendey, B., 2012. Evidence for angiogenesis in Parkinson's disease, incidental Lewy body disease, and progressive supranuclear palsy. J. Neural Transm. 119, 59–71. Vienna Austria.

Dillon, C.D., Billings, M., Hockey, K.S., DeLaGarza, L., Rzigalinski, B.A., 2011. Cerium oxide nanoparticles protect against MPTP-induced dopaminergic neurodegeneration in a mouse model for Parkinson's disease. NSTI-Nanotech. 3, 451–454.

Duarte, J., Perez-Vizcaino, F., Zarzuelo, A., Jimenez, J., Tamargo, J., 1993. Vasodilator effects of quercetin in isolated rat vascular smooth muscle. Eur. J. Pharmacol. 239, 1–7.

Enteshari Najafabadi, R., Kazemipour, N., Esmaeili, A., Beheshti, S., Nazifi, S., 2018. Using superparamagnetic iron oxide nanoparticles to enhance bioavailability of quercetin in the intact rat brain. BMC Pharmacol. Toxicol. 19, 59.

Erdo, F., Denes, L., de Lange, E., 2017. Age-associated physiological and pathological changes at the blood-brain barrier: a review. J. Cereb. Blood Flow Metab. 37, 4–24.

Estevez, A.Y., Erlichman, J.S., 2014. The potential of cerium oxide nanoparticles for neurodegenerative disease therapy. Nanomedicine 9, 1437–1440.

Gandhi, S., Abramov, A.Y., 2012. Mechanisms of oxidative stress in neurodegeneration. Oxid. Med. Cell Longev. 2012, 428010.

Grabrucker, A.M., Ruozi, B., Belletti, D., Pederzoli, F., Forni, F., Vandelli, M.A., et al., 2016. Nanoparticle transport across the blood brain barrier. Tissue Barriers 4, e1153568.

Gonzalez-Carter, D.A., Ong, Z.Y., McGilvery, C.M., Dunlop, I.E., Dexter, D.T., Porter, A.E., 2018. L-DOPA functionalized, multi-branched gold nanoparticles as brain-targeted nano-vehicles. Nanomedicine 15, 1–11.

Goode, A.E., Gonzalez-Carter, D.A., Motskin, M., Pienaar, I., Chen, S., Hu, S., et al., 2015. High resolution and dynamic imaging of biopersistence and bioreactivity of extra and intracellular MWNTs exposed to microglial cells. Biomaterials 70, 57—70.

Graham, U.M., Tseng, M.T., Jasinski, J.B., Yokel, R.A., Unrine, J.M., Davis, B.H., et al., 2014. *In vivo* processing of ceria nanoparticles inside liver: impact on free-radical scavenging activity and oxidative stress. ChemPlusChem 79, 1083—1088.

Gray, M.T., Woulfe, J.M., 2015. Striatal blood-brain barrier permeability in Parkinson's disease. J. Cereb. Blood Flow Metab. Off. J. Int. Soc. Cereb. Blood Flow Metab. 35, 747—750.

Gromnicova, R., Davies, H.A., Sreekanth reddy, P., Romero, I.A., Lund, T., Roitt, I.M., et al., 2013. Glucose-coated gold nanoparticles transfer across human brain endothelium and enter astrocytes *in vitro*. PLoS One 8, e81043.

Gupta, A., Das, S., Seal, S., 2014. Redox-active nanoparticles in combating neurodegeneration. Nanomedicine 9, 2725—2728.

Hans, M.L., Lowman, A.M., 2002. Biodegradable nanoparticles for drug delivery and targeting. Curr. Opin. Solid State Mater. Sci. 6, 319—327.

Heckman, K.L., DeCoteau, W., Estevez, A., Reed, K.J., Costanzo, W., Sanford, D., et al., 2013. Custom cerium oxide nanoparticles protect against a free radical mediated autoimmune degenerative disease in the brain. ACS Nano 7, 10582—10596.

Hsieh, H.L., Yang, C.M., 2013. Role of redox signaling in neuroinflammation and neurodegenerative diseases. BioMed Res. Int. 484613.

Hu, K., Shi, Y., Jiang, W., Han, J., Huang, S., Jiang, X., 2011. Lactoferrin conjugated PEG-PLGA nanoparticles for brain delivery: preparation, characterization and efficacy in Parkinson's disease. Int. J. Pharm. 415, 273—283.

Huang, Y., Zhang, B., Xie, S., Yang, B., Xu, Q., Tan, J., 2016. Superparamagnetic Iron oxide nanoparticles modified with tween 80 pass through the intact blood—brain barrier in rats under magnetic field. ACS Appl. Mater. Interfaces 8, 11336—11341.

Hu, K., Chen, X., Chen, W., Zhang, L., Li, J., Ye, J., et al., 2018. Neuroprotective effect of gold nanoparticles composites in Parkinson's disease model. Nanomedicine 14, 1123—1136.

Iadecola, C., 2013. The pathobiology of vascular dementia. Neuron 80, 844—866.

Iadecola, C., 2004. Neurovascular regulation in the normal brain and in Alzheimer's disease. Nat. Rev. Neurosci. 5, 347—360.

Janelidze, S., Hertze, J., Nagga, K., Nilsson, K., Nilsson, C., , Swedish BioFINDER Study Group, Wennstrom, M., et al., 2017. Increased blood-brain barrier permeability is associated with dementia and diabetes but not amyloid pathology or APOE genotype. Neurobiol. Aging 51, 104—112.

Jiang, X., Xin, H., Ren, Q., Gu, J., Zhu, L., Du, F., et al., 2014. Nanoparticles of 2-deoxy-d-glucose functionalized poly(ethylene glycol)-*co*-poly(trimethylene carbonate) for dual-targeted drug delivery in glioma treatment. Biomaterials 35, 518—529.

Johnston, T.H., Fox, S.H., Brotchie, J.M., 2005. Advances in the delivery of treatments for Parkinson's disease. Expert Opin. Drug Deliv. 2, 1059—1073.

Juillerat-Jeanneret, L., 2008. The targeted delivery of cancer drugs across the blood-brain barrier: chemical modifications of drugs or drug-nanoparticles? Drug Discov. Today 13, 1099—1106.

Kang, Y.J., Cutler, E.G., Cho, H., 2018. Therapeutic nanoplatforms and delivery strategies for neurological disorders. Nano Converg. 5, 35.

Katzenschlager, R., Hughes, A., Evans, A., Manson, A.J., Hoffman, M., Swinn, L., et al., 2005. Continuous subcutaneous apomorphine therapy improves dyskinesias in Parkinson's disease: a prospective study using single-dose challenges. Mov. Disord. 20, 151—157.

Khan, A.R., Yang, X., Fu, M., Zhai, G., 2018. Recent progress of drug nanoformulations targeting to brain. J. Control. Release 291, 37–64.

Kisler, K., Nelson, A.R., Montagne, A., Zlokovic, B.V., 2017. Cerebral blood flow regulation and neurovascular dysfunction in Alzheimer disease. Nat. Rev. Neurosci. 18, 419–434.

Kulkarni, P.V., Roney, C.A., Antich, P.P., Bonte, F.J., Raghu, A.V., 2010. Quinoline-n-butylcyanoacrylate-based nanoparticles for brain targeting for the diagnosis of Alzheimer's disease. Wiley Interdiscip. Rev. Nanomed. Nanobiotechnol. 2, 35–47.

Kumar, A., Das, S., Munusamy, P., Self, W., Baer, D.R., Sayle, D.C., et al., 2014. Behavior of nanoceria in biologically-relevant environments. Environ. Sci. Nano. 1, 516–532.

Laurent, S., Burtea, C., Thirifays, C., Hafeli, U.O., Mahmoudi, M., 2012. Crucial ignored parameters on nanotoxicology: the importance of toxicity assay modifications and "cell vision". PLoS One 7, e29997.

Lee, E.S., Youn, Y.S., 2016. Albumin-based potential drugs: focus on half-life extension and nanoparticle preparation. J. Pharm. Invest. 46, 305–315.

Li, W., Luo, R., Lin, X., et al., 2015. Remote modulation of neural activities via near-infrared triggered release of biomolecules. Biomaterials 65, 76–85.

Loeffler, D.A., Connor, J.R., Juneau, P.L., Snyder, B.S., Kanaley, L., DeMaggio, A.J., et al., 1995. Transferrin and iron in normal, Alzheimer's disease, and Parkinson's disease brain regions. J. Neurochem. 65, 710–724.

Mahmoudi, M., Laurent, S., Shokrgozar, M.A., Hosseinkhani, M., 2011. Toxicity evaluations of superparamagnetic iron oxide nanoparticles: cell "vision" versus physicochemical properties of nanoparticles. ACS Nano 5, 7263–7276.

Mena, M.A., Garcia de Yebenes, J., 2008. Glial cells as players in parkinsonism: the "good," the "bad," and the "mysterious" glia. The Neuroscientist 14, 544–560.

Miyakawa, T., 2010. Vascular pathology in Alzheimer's disease. Psychogeriatrics 10, 39–44.

Muller, R.H., Jacobs, C., Kayser, O., 2001. Nanosuspensions as particulate drug formulations in therapy: rationale for development and what we can expect for the future. Adv. Drug Deliv. Rev. 47, 3–19.

Naz, S., Beach, J., Heckert, B., Tummala, T., Pashchenko, O., Banerjee, T., et al., 2017. Cerium oxide nanoparticles: a 'radical' approach to neurodegenerative disease treatment. Nanomedicine (Lond) 12, 545–553.

Nazem, A., Mansoori, G.A., 2011. Nanotechnology for Alzheimer's disease detection and treatment. Insci. J. 1, 169–193.

Neuwelt, E., Abbott, N.J., Abrey, L., Banks, W.A., Blakley, B., Davis, T., et al., 2008. Strategies to advance translational research into brain barriers. Lancet Neurol. 7, 84–96.

Nyholm, D., Nilsson Remahl, A.I., Dizdar, N., Constantinescu, R., Holmberg, B., Jansson, R., et al., 2005. Duodenal levodopa infusion monotherapy vs oral polypharmacy in advanced Parkinson disease. Neurology 64, 216–223.

Obulesu., M., 2019. Chapter 5. Blood brain barrier targeted nanotechnological advances Obulesu M.Alzheimer's Disease Theranostics. Elsevier, p. 253.

Obeso, J.A., Rodriguez-Oroz, M.C., Goetz, C.G., Marin, C., Kordower, J.H., Rodriguez, M., et al., 2010. Missing pieces in the Parkinson's disease puzzle. Nat. Med. 16, 653–661.

Obulays, M., 2018. Chapter 16. Multifarious therapeutic avenues for Alzheimer's disease. In: Singh, S. (Ed.), Pathology, Prevention and Therapeutics of Neurodegenerative Disease. Springer, pp. 185–190.

Olivier, J.C., 2005. Drug transport to brain with targeted nanoparticles. NeuroRx 2, 108–119.

Orr, C.F., Rowe, D.B., Halliday, G.M., 2002. An inflammatory review of Parkinson's disease. Prog. Neurobiol. 68, 325–340.

Patching, S.G., 2017. Glucose transporters at the blood—brain barrier: function, regulation and gateways for drug delivery. Mol. Neurobiol. 54, 1046—1077.

Persidsky, Y., Ramirez, S.H., Haorah, J., Kanmogne, G.D., 2006. Blood-brain barrier: structural components and function under physiologic and pathologic conditions. J. Neuroimmune Pharmacol. 1, 223—236.

Pienaar, I.S., Lee, C.H., Elson, J.L., McGuinness, L., Gentleman, S.M., Kalaria, R.N., et al., 2015. Deep-brain stimulation associates with improved microvascular integrity in the subthalamic nucleus in Parkinson's disease. Neurobiol. Dis. 74, 392—405.

Pinna, A., Malfatti, L., Galleri, G., Manetti, R., Cossu, S., Rocchitta, G., et al., 2015. Ceria nanoparticles for the treatment of Parkinson-like diseases induced by chronic manganese intoxication. RSC Adv. 5, 20432—20439.

Pisani, V., Stefani, A., Pierantozzi, M., Natoli, S., Stanzione, P., Franciotta, D., et al., 2012. Increased blood-cerebrospinal fluid transfer of albumin in advanced Parkinson's disease. J. Neuroinflammation 9, 188.

Popovtzer, R., Agrawal, A., Kotov, N., Popovtzer, A., Balter, J., Carey, T.E., et al., 2008. Targeted gold nanoparticles enable molecular CT imaging of cancer. Nano Lett. 8, 4593—4596.

Portioli, C., Benati, D., Pii, Y., Bernardi, P., Crucianelli, M., Santucci, S., et al., 2013. Short-term biodistribution of cerium oxide nanoparticles in mice: focus on brain parenchyma. Nanosci. Nanotechnol. Lett. 5, 1174—1181.

Pulido-Reyes, G., Rodea-Palomares, I., Das, S., Sakthivel, T.S., Leganes, F., Rosal, R., et al., 2015. Untangling the biological effects of cerium oxide nanoparticles: the role of surface valence states. Sci. Rep. 5, 15613.

Ramos, F.A., Takaishi, Y., Shirotori, M., Kawaguchi, Y., Tsuchiya, K., Shibata, H., et al., 2006. Antibacterial and antioxidant activities of quercetin oxidation products from yellow onion (Allium cepa) skin. J. Agric. Food Chem. 54, 3551—3557.

Ramanathan, S., Archunan, G., Sivakumar, M., Tamil Selvan, S., Fred, A.L., Kumar, S., et al., 2018. Theranostic applications of nanoparticles in neurodegenerative disorders. Int. J. Nanomed. 13, 5561—5576.

Rojas, S., Gispert, J.D., Abad, S., Buaki-Sogo, M., Victor, V.M., Garcia, H., et al., 2012. *In vivo* biodistribution of amino-functionalized ceria nanoparticles in rats using positron emission tomography. Mol. Pharm. 9, 3543—3550.

Shilo, M., Motiei, M., Hana, P., Popovtzer, R., 2014. Transport of nanoparticles through the blood-brain barrier for imaging and therapeutic applications. Nanoscale 6, 2146—2152.

Stewart, L.K., Soileau, J.L., Ribnicky, D., Wang, Z.Q., Raskin, I., Poulev, A., et al., 2008. Quercetin transiently increases energy expenditure but persistently decreases circulating markers of inflammation in C57BL/6J mice fed a high fat diet. Metabolism 57, S39—S46.

Sweeney, M.D., Sagare, A.P., Zlokovic, B.V., 2018. Blood-brain barrier breakdown in Alzheimer disease and other neurodegenerative disorders. Nat. Rev. Neurol. 14, 133—150.

Sweeney, M.D., Ayyadurai, S., Zlokovic, B.V., 2016. Pericytes of the neurovascular unit: key functions and signaling pathways. Nat. Neurosci. 19, 771—783.

Valdiglesias, V., Kilic, G., Costa, C., Fernandez-Bertolez, N., Pasaro, E., Teixeira, J.P., et al., 2015. Effects of iron oxide nanoparticles: cytotoxicity, genotoxicity, developmental toxicity, and neurotoxicity. Environ. Mol. Mutagen. 56, 125—148.

Vinzant, N., Scholl, J.L., Wu, C.M., Kindle, T., Koodali, R., Forster, G.L., 2017. Iron oxide nanoparticle delivery of peptides to the brain: reversal of anxiety during drug withdrawal. Front. Neurosci. 11, 608.

Wada, K., Arai, H., Takanashi, M., Fukae, J., Oizumi, H., Yasuda, T., et al., 2006. Expression levels of vascular endothelial growth factor and its receptors in Parkinson's disease. Neuroreport 17, 705–709.

Walkey, C., Das, S., Seal, S., Erlichman, J., Heckman, K., Ghibelli, L., et al., 2015. Catalytic properties and biomedical applications of cerium oxide nanoparticles. Environ. Sci. Nano. 2, 33–53.

Wiley, D.T., Webster, P., Gale, A., Davis, M.E., 2013. Transcytosis and brain uptake of transferrin-containing nanoparticles by tuning avidity to transferrin receptor. Proc. Natl. Acad. Sci. U.S.A. 110, 8662–8667.

Wilson, B., Samanta, M.K., Santhi, K., Kumar, K.P.S., Paramakrishnan, N., Suresh, B., 2008. Targeted delivery of tacrine into the brain with polysorbate 80-coated poly(n-butylcyanoacrylate) nanoparticles. Eur. J. Pharm. Biopharm. 70, 75–84.

Wilson, B., Samanta, M.K., Santhi, K., Kumar, K.P.S., Ramasamy, M., Suresh, B., 2010. Chitosan nanoparticles as a new delivery system for the anti-Alzheimer drug tacrine. Nanomedicine 6, 744–750.

Xu, Y., Hun, X., Liu, F., Wen, X., Luo, X., 2015. Aptamer biosensor for dopamine based on a gold electrode modified with carbon nanoparticles and thionine labeled gold nanoparticles as probe. Microchim. Acta 182, 1797–1802.

Yokel, R.A., Tseng, M.T., Dan, M., Unrine, J.M., Graham, U.M., Wu, P., et al., 2013. Biodistribution and biopersistence of ceria engineered nanomaterials: size dependence. Nanomedicine 9, 398–407.

Yurek, D.M., Flectcher, A.M., Kowalczyk, T.H., Padegimas, L., Cooper, M.J., 2009. Compacted DNA nanoparticle gene transfer of GDNF to the rat striatum enhances the survival of grafted fetal dopamine neurons. Cell Transplant. 18, 1183–1196.

Zlokovic, B.V., 2011. Neurovascular pathways to neurodegeneration in Alzheimer's disease and other disorders. Nat. Rev. Neurosci. 12, 723–738.

Zlokovic, B.V., 2008. The blood-brain barrier in health and chronic neurodegenerative disorders. Neuron 57, 178–201.

Further reading

Obulesu, M., 2019. Chapter 5. Blood brain barrier targeted nanotechnological advances. In: Obulesu, M. (Ed.), Alzheimer's Disease Theranostics. Elsevier, pp. 25–32.

CHAPTER 4

Natural compounds in the treatment of Parkinson's disease

1. Introduction

1.1 Oxidative stress

Enhanced reactive oxygen species (ROS) production has been implicated in the etiology of numerous diseases such as neurodegenerative diseases (Ding et al., 2018). Primarily, oxidative stress is exerted because of the inequality in prooxidants and antioxidants in cells (Barnham et al., 2004; Ding et al., 2018). Essentially, plethora of ROS is produced in the cell during the synthesis of adenosine triphosphate (ATP) in mitochondria (Shadel and Horvath, 2015; Ding et al., 2018). Dopaminergic (DA) neurons require enormous amount of ATP to produce and liberate dopamine (Mamelak, 2018; Ding et al., 2018). Accordingly, higher amount of ROS are generated in DA neurons than other neurons (Ding et al., 2018). In addition, enhanced ROS production takes place during the aging process because of the growth of mitochondrial DNA (mtDNA) mutations (Gredilla et al., 2012; Ding et al., 2018). Generally, production and elimination of ROS is perfectly mediated by cellular antioxidative enzymes such as superoxide dismutase (SOD) and catalase (Ding et al., 2018). In Parkinson's disease (PD), mitochondrial impairment, aging, iron aggregation, dopamine metabolism, and increased sensitivity of aged neurons to oxidative stress are vital events contributing for enhanced oxidative stress (Floyd and Carney, 1992; Ding et al., 2018). While the molecular underpinnings involved in etiopathogenesis of PD are hitherto unresolved issues, oxidative stress has been found to be the major cause of the disease (Jenner, 2003; Przedborski, 2017; Ding et al., 2018).

Parkinson's Disease Therapeutics
ISBN 978-0-12-819882-7
https://doi.org/10.1016/B978-0-12-819882-7.00004-0

2. Plant-derived compounds

Neuroinflammation plays a pivotal role in neurodegenerative diseases like Alzheimer's disease (AD), PD (Moore et al., 2010; Lull and Block, 2010; Filiou et al., 2014; Obulesu and Jhansilakshmi, 2014; Clark and Kodadek, 2016; Kim et al., 2018). Commonly, plants, animals, and microorganisms produce manifold metabolites, which primarily belong to the class quinolones (Heeb et al., 2011; Kim et al., 2018). In line with this, 4-hydroxy-2-alkylquinoline (pseudane-VII) isolated from *Pseudoalteromonas* sp. M2 in marine water has been found to ameliorate lipopolysaccharide-induced neuroinflammation by decreasing nitric oxide (NO) and ROS generation and the expression of inducible nitric oxide synthase (iNOS) and cyclooxygenase-2 (COX-2) in both in vitro and in vivo (Kim et al., 2017, 2018).

2.1 Chinese herbs

Chinese herbs, which have long been used in the treatment of multifarious diseases in China, were found to exhibit antiinflammatory, antioxidant, antiapoptosis, free radical removal, and hazardous metal chelating effects (Fu et al., 2015; Ding et al., 2018). Many lines of evidence have also indicated that these herbs showed significantly low toxic side effects (Ding et al., 2018). As Chinese herbs are abundant in antioxidants, they show remarkably better therapeutic efficacy in PD via multifarious biologicals pathways (Obrenovich et al., 2010; Soobrattee et al., 2010; Ding et al., 2018).

2.2 Flavonoids

Flavonoids chiefly present in plethora of plants, fruits, and vegetables render valuable pharmacological effects (Liu et al., 2014; Nabavi et al., 2015; de Andrade Teles et al., 2018). Additionally, they exhibit potential blood—brain barrier (BBB) spanning ability, thus showing protective efficacy against neurodegenerative diseases such as AD and PD (Elbaz et al., 2016; de Andrade Teles et al., 2018). Kaempferol, an essential flavonol of tea, broccoli, Brussel sprouts, apple, and grape fruit improved SOD and Gpx activity in the mouse model of PD by its robust antioxidant and antiinflammatory characteristics (Li and Pu, 2011; Zuk et al., 2011; Ding et al., 2018).

2.3 Mitochondria

2.3.1 Epigallocatechin gallate

Epigallocatechin gallate (EGCG), a vital ingredient of green tea, exhibits antioxidant, antiapoptotic, and free radical elimination activity. Accordingly, it showed upregulation of peroxisome proliferator-activated receptor gamma coactivator 1 alpha (PGC-1α), thus ameliorating mitochondrial function and DA neuronal longevity (Ye et al., 2012; Ding et al., 2018). Additionally, EGCG has been found to attenuate mitochondrial impairment in mutant leucine-rich repeat kinase 2 and parkin null flies via initiation of adenosine monophosphate—activated protein kinase signaling pathway (Ng et al., 2012; Ding et al., 2018). DL-3-n-butylphthalide obtained from 1,3-n-butylphthalide isolated from the seeds of *Apium graveolens* Linn. (Chinese celery) has been identified to scavenge ROS, thus restoring mitochondrial function in an 1-methyl-4-phenyl-pyridinium(MPP^+)-treated cellular PD model (Li et al., 2009; Huang et al., 2010; Ding et al., 2018).

2.4 Silibinin

Silibinin, (2R,3R)-3,5,7-trihydroxy-2-[(2R, 3R)-3-(4-hydroxy-3-methoxy phenyl)-2-(hydroxymethyl)-2,3-dihydro-1,4-benzodioxin-6-yl]-2,3-dihydro-4H-chromen-4-one, is a flavonoid isolated from the species *Silybum marianum* (Svagera et al., 2003; de Andrade Teles et al., 2018). Mounting evidence has shown the protective efficacy of spatial memory, leaning, and locomotor activity through its antiinflammatory and antioxidant activity (Lim et al., 2014; Wang et al., 2016; Song et al., 2017; de Andrade Teles et al., 2018). It has also been reported that silibinin restores dopamine levels in the striatum and increases glutathione (GSH) and decreases malondialdehyde in aging animals, thus showing protective efficacy against PD (Chen et al., 2015a,b; de Andrade Teles et al., 2018). Additionally, Lee et al. (2015) also emphasized the neuroprotective mechanism of silibinin against 1-methyl-4-phenyl-1,2,3,6-tetrahydropyridine(MPTP) PD model by averting motor dysfunction and neuronal loss (de Andrade Teles et al., 2018).

2.5 Naringenin

Naringenin, (5,7-dihydroxy-2-(4-hydroxyphenyl)-chroman-4-one), a flavonoid of citrus fruits, is abundantly available in grapes (de Andrade Teles et al., 2018). Abundance of evidence has shown the antioxidant, antiinflammatory, and antinociceptive effects of naringenin (Manchope et al., 2016; de Andrade Teles et al., 2018). Accordingly, studies have shown the protective efficacy of naringenin against 6-OHDA induced neurodegeneration in PD model (de Andrade Teles et al., 2018). It improved nuclear factor E2-related factor 2 (Nrf2) (mediating factor of the antioxidant protein gene expression) and initiated antioxidant response element pathway genes, in turn mitigating ROS production in the striatum (Lou et al., 2014; de Andrade Teles et al., 2018).

2.6 Baicalein

Baicalein (5,6,7-trihydroxyflavone), a flavonoid extracted from roots of *Scutellaria baicalensis* and *Scutellaria lateriflora*, has antioxidant, antiinflammatory, and neuroprotective properties (Li et al., 2005; Makino et al., 2008; de Andrade Teles et al., 2018). Studies have also shown that baicalein suppressed motor deficits, enhanced neurotransmitters such as dopamine (DA), DOPAC (3,4-dihydroxyphenylacetic acid), and HVA (homovanillic acid), averted dopaminergic neuron loss, and reduced α-syn oligomers (Lee et al., 2014; Hu et al., 2016; de Andrade Teles et al., 2018). It also reduces the toxicity of MPP, which adversely affects the oxidative phosphorylation in mitochondria (Lee et al., 2014). Moreover, pretreatment with baicalein decreased microglial and astrocyte activation (de Andrade Teles et al., 2018).

2.7 Polyphenols

Resveratrol, a polyphenol abundantly found in grapes, exerted protective efficacy by ameliorating mitochondrial impairment via perturbing energy metabolic sensors by regulating autophagy signals and initiation of nicotinamide adenine nucleotide—dependent deacetylase sirtuin-1 (SIRT1) and PGC-1α (Lagouge et al., 2006; Jeong et al., 2012; Ding et al., 2018). In addition, it also improved mRNA levels of numerous PGC-1α genes such as mitochondrial transcription factor A (TFAM) and cyclooxygenase-1 (COX-1), thus facilitating mitochondrial biogenesis (Ferretta et al., 2014; Ding et al., 2018). In addition, polydatin, a glycosylated form of resveratrol, ameliorated GSH and manganese superoxide dismutases, which were reduced in rotenone-treated rodent model of PD (Ding et al., 2018).

Similarly, chrysotoxine, an essential bioactive bibenzyl compound extracted from medicinal *Dendrobium* species with adequate ROS scavenging activity, showed protection against 6-hydroxydopamine (6-OHDA) instigated intracellular production of ROS and mitochondrial dysfunction such as reduction of mitochondrial membrane potential (MMP), augmentation of intracellular free Ca^{2+}, discharge of cytochrome C, and inequality of Bax/Bcl-2 ratio (Zhang et al., 2007; Song et al., 2010; Ding et al., 2018). Morin and mangiferin are two vital polyphenols chiefly and extensively found in vegetables, fruits, plant extracts, tea, and wine and were also observed to decrease ROS levels by reinstating MMP in excitotoxic instigated cell model (Campos-Esparza et al., 2009; Kavitha et al., 2014; Ding et al., 2018). Salidroside, a phenylpropanoid glycoside extracted from *Rhodiola rosea* L., combated MPP + instigated toxicity by decreasing ROS or nitric oxide (NO) generation, reducing cytochrome *c* and Smac release, mediating Bax/Bcl2 ratio, and attenuating caspase activation in DA neurons (Wang et al., 2015; Ding et al., 2018). Paeonol, a pivotal polyphenolic ingredient of Chinese herb, *Cortex Moutan*, with adequate antioxidant, antiinflammatory, and antitumor activities showed significant inhibition of intracellular ROS and mitochondrial cell death pathways such as MMP disintegration, caspase 3 activation, and cytochrome *c* release in MPP + induced cellular PD model (Lu et al., 2015; Ding et al., 2018).

2.8 Polyphenolic flavonoids

2.8.1 Anthocyanins

Anthocyanins are polyphenolic flavonoids abundantly available in vegetables, grains, fruits, and flowers (Wang et al., 2008; de Andrade Teles et al., 2018). They show potential antioxidant, antiinflammatory, and antiapoptotic characteristics and play a role in ameliorating cognition and memory (Winter et al., 2017; de Andrade Teles et al., 2018). Anthocyanin molecule comprises 3 components such as an aglycone (anthocyanidin), a group of sugars, and frequently a group of organic acids (Smeriglio et al., 2016; de Andrade Teles et al., 2018). These compounds were conjugated to gold nanoparticles also to improve memory in AD mouse models (Alim et al., 2016; de Andrade Teles et al., 2018).

3. Amelioration of antioxidant enzymes

Cells comprise enzymatic and nonenzymatic antioxidant protective systems (Gilgun-Sherki et al., 2001; Ding et al., 2018). However, with aging and under disease conditions, these systems have been found to deteriorate (Sohal and Weindruch, 1996; Lotharius and O'Malley, 2000; Ding et al., 2018). Polyphenols with plethora of protective effects also play a pivotal role in enhancing the antioxidative enzyme activity (Ebrahimi and Schluesener, 2012; Ding et al., 2018).

Gypenosides (GP), saponins isolated from *Gynostemma pentaphyllum* treatment, averted MPTP-provoked reduction of GSH and decreased SOD activity in the SNpc of the mice (Wang et al., 2010; Ding et al., 2018). Quercetin being the vital ingredient of vegetables and fruits, olive oil, onions, and red wine enhanced GSH levels in striatum of 6-OHDA-treated rats (Miean and Mohamed, 2001; Haleagrahara et al., 2011; Ding et al., 2018). A sesquiterpene alcohol namely nerolidol profoundly enhanced the levels of SOD, catalase, and GSH in rotenone-instigated PD model (Javed et al., 2016; Ding et al., 2018).

3.1 Glutathione

Augmenting the levels of GSH, a tripeptide has been found to deteriorate intracellular ROS levels, thus protecting neurons in PD (Aaseth et al., 2018). In line with this, wealth of studies has also shown that GSH levels are low in substantia nigra(SN) unlike in other brain regions preferably in early stages of PD due to the negative impact of extravesicular DA and its degradation products on GSH levels (Pearce et al., 1997; Stokes et al., 1999, Aaseth et al., 2018). GSH acts by itself or in combination with GSH peroxidase enzyme to mitigate ROS (Meister and Anderson 1983; Aaseth et al., 2018). In addition, N-acetylcysteine was observed to improve GSH levels in MPTP or rotenone-treated PD models by preventing DA decrease in SN (Chinta et al., 2006; Rahimmi et al., 2015; Aaseth et al., 2018).

3.2 Metal chelation

Iron aggregated in the aging brain followed by misfolded protein accumulation has been found to induce neurodegeneration in both AD and PD (Obulesu et al., 2011; Devos et al., 2014; Ward et al., 2014; Aaseth et al., 2018; Ding et al., 2018). Several lines of evidence have shown that iron and copper augment oxidative stress and promote PD progression (Ward et al., 2014; Dusek et al., 2015; Genoud et al., 2017; Aaseth et al., 2018). Iron increases ROS production by Fenton reaction and Haber—Weiss reaction (Barnham et al., 2004; Ding et al., 2018). Ginsenoside, a vital ingredient extracted from ginseng, mitigated the 6-OHDA instigated iron intake by attenuating the upregulation of an iron transferring protein with its substantial antioxidant effect (Xu et al., 2010; Ding et al., 2018). EGCG also acted as substantial iron chelator and suppressed the transition metal initiated free radicals, thus exhibiting its antioxidant protective effects (Weinreb et al., 2009; Singh et al., 2016; Ding et al., 2018). In addition, it also mitigates iron aggregation in PD by perturbing iron-responsive element (Reznichenko et al., 2006; Ding et al., 2018). A few iron binding molecules such as nicotine, niacin, caffeine, and salbutamol showed significant reduction in PD progress (Aaseth et al., 2018).

Caffeine treatment showed significant improvement in PD symptoms in a small group of patients (Postuma et al., 2012; Aaseth et al., 2018). However, similar results could not be obtained when the same treatment was used in a larger multicentre study (Postuma et al., 2017; Aaseth et al., 2018). The protective efficacy of caffeine is due to its iron binding ability through its nitrogen and oxygen group with lower binding constant than that of ethylenediaminetetraacetate (EDTA)-Fe (Andjelkovic et al., 2006; Aaseth et al., 2018). Vitamin B3 or niacin with iron binding ability also shows protective efficacy in PD by improving dopamine synthesis and decreasing inflammation (Al-Saif and Refat, 2012; Aaseth et al., 2018). Nicotine with its substantial iron binding ability also shows neuroprotective efficacy against PD (Zhang et al., 2006; Fazary, 2017; Aaseth et al., 2018).

4. Conclusions and future perspectives

Several Chinese herbs show ample of therapeutic benefits such as antioxidant and antiinflammatory properties. Therefore, there are ample of potential molecules rendering valuable therapeutic properties in multifarious pathways primarily by scavenging ROS and exerting antioxidative properties (Song et al., 2012; Ding et al., 2018). However, further studies on evaluation of the BBB spanning ability and therapeutic ability of these molecules against PD are warranted. Only similar lines and several flavonoids such as EGCG, silibinin, and baicalein exhibit impressive antioxidant, antiinflammatory, and neuroprotective properties. Polyphenols such as resveratrol, polydatin, paeonol, morin, and mangiferin also showed similar antioxidant, antiinflammatory, and neuroprotective properties. Polyphenolic flavonoids such as anthocyanins also showed exemplary neuroprotective properties. Despite the remarkable therapeutic efficacy of these compounds, further studies are warranted to validate their efficacy. In addition, their bioavailability and short systemic circulation may impede their success. Therefore, multifarious drug delivery systems may be required to overcome these challenges.

References

Aaseth, J., Dusek, P., Roos, P.M., 2018. Prevention of progression in Parkinson's disease. Biometals 31, 737–747.

Alim, T., Kim, M.J., Rehman, S.U., Ahmad, A., Kim, M.O., 2016. Anthocyanin-loaded PEG-gold nanoparticles enhanced the neuroprotection of anthocyanins in an Aβ1-42 mouse model of Alzheimer's disease. Mol. Neurobiol. 54, 6490–6506.

Al-Saif, F.A., Refat, M.S., 2012. Ten metal complexes of vitamin B 3/niacin: spectroscopic, thermal, antibacterial, antifungal, cytotoxicity and antitumor studies of Mn(II), Fe(III), Co(II), Ni(II), Cu(II), Zn(II), Pd(II), Cd(II), Pt(IV) and Au(III) complexes. J. Mol. Struct. 1021, 40–52.

Andjelkovic, M., Van Camp, J., De Meulenaer, B., Depaemelaere, G., Socaciu, C., Verloo, M., et al., 2006. Iron-chelation properties of phenolic acids bearing catechol and galloyl groups. Food Chem. 98, 23–31.

Barnham, K.J., Masters, C.L., Bush, A.I., 2004. Neurodegenerative diseases and oxidative stress. Nat. Rev. Drug Discov. 3, 205–214.

Campos-Esparza, M.R., Sanchez-Gomez, M.V., Matute, C., 2009. Molecular mechanisms of neuroprotection by two natural antioxidant polyphenols. Cell Calcium 45, 358–368.

Chen, Y., Zhang, D.Q., Liao, Z., Wang, B., Gong, S., Wang, C., et al., 2015a. Antioxidant polydatin (piceid) protects against substantia nigral motor degeneration in multiple rodent models of Parkinson's disease. Mol. Neurodegener. 10, 4.

Chen, H., Wang, X., Wang, M., Yang, L., Yan, Z., Zhang, Y., et al., 2015b. Behavioral and neurochemical deficits in aging rats with increased neonatal iron intake: silibinin's neuroprotection by maintaining redox balance. Front. Aging Neurosci. 206.

Chinta, S.J., Kumar, J.M., Zhang, H., Forman, H.J., Andersen, J.K., 2006. Up-regulation of gamma-glutamyl transpeptidase activity following glutathion depletion has a compensatory rather than an inhibitory effect on mitochondrial complex I activity: implications for Parkinson's disease. Free Radic. Biol. Med. 40, 1557–1563.

Clark, L.F., Kodadek, T., 2016. The immune system and neuroinflammation as potential sources of blood-based biomarkers for Alzheimer's disease, Parkinson's disease, and Huntington's disease. ACS Chem. Neurosci. 7, 520–527.

de Andrade Teles, R.B., Diniz, T.C., Costa Pinto, T.C., de Oliveira Junior, R.G., Gama E Silva, M., de Lavor, E.M., et al., 2018. Flavonoids as therapeutic agents in Alzheimer's and Parkinson's diseases: a systematic review of preclinical evidences. Oxid. Med. Cell Longev. 7043213.

Devos, D., Moreau, C., Devedjian, J.C., Kluza, J., Petrault, M., Laloux, C., et al., 2014. Targeting chelatable iron as a therapeutic modality in Parkinson's disease. Antioxidants Redox Signal. 2, 195–210.

Ding, Y., Xin, C., Zhang, C.W., Lim, K.L., Zhang, H., Fu, Z., et al., 2018. Natural molecules from Chinese herbs protecting against Parkinson's disease via anti-oxidative stress. Front. Aging Neurosci. 10, 246.

Dusek, P., Roos, P.M., Litwin, T., Schneider, S.A., Flaten, T.P., Aaseth, J., 2015. The neurotoxicity of iron, copper and manganese in Parkinson's and Wilson's diseases. J. Trace Elem. Med. Biol. 31, 193–203.

Ebrahimi, A., Schluesener, H., 2012. Natural polyphenols against neurodegenerative disorders: potentials and pitfalls. Ageing Res. Rev. 11, 329–345.

Elbaz, A., Carcaillon, L., Kab, S., Moisan, F., 2016. Epidemiology of Parkinson's disease. Rev. Neurol. 172, 14–26.

Fazary, A.E., 2017. Metal Complexes of Nicotine: A Group of Negligible Compounds, vol. 8. Pelagia Research Library, Abha.

Filiou, M.D., Arefin, A.S., Moscato, P., Graeber, M.B., 2014. Neuroinflammation' differs categorically from inflammation: transcriptomes of Alzheimer's disease, Parkinson's disease, schizophrenia and inflammatory diseases compared. Neurogenetics 15, 201–212.

Floyd, R.A., Carney, J.M., 1992. Free radical damage to protein and DNA: mechanisms involved and relevant observations on brain undergoing oxidative stress. Ann. Neurol. 32, 22–27.

Fu, W., Zhuang, W., Zhou, S., Wang, X., 2015. Plant-derived neuroprotective agents in Parkinson's disease. Am. J. Transl. Res. 7, 1189–1202.

Ferretta, A., Gaballo, A., Tanzarella, P., Piccoli, C., Capitanio, N., Nico, B., et al., 2014. Effect of resveratrol on mitochondrial function: implications in parkin associated familiar Parkinson's disease. Biochim. Biophys. Acta 1842, 902–915.

Genoud, S., Roberts, B.R., Gunn, A.P., Halliday, G.M., Lewis, S.J.G., Ball, H.J., et al., 2017. Subcellular compartmentalisation of copper, iron, manganese, and zinc in the Parkinson's disease brain. Metallomics 9, 1447–1455.

Gilgun-Sherki, Y., Melamed, E., Offen, D., 2001. Oxidative stress induced neurodegenerative diseases: the need for antioxidants that penetrate the blood brain barrier. Neuropharmacology 40, 959–975.

Gredilla, R., Weissman, L., Yang, J.L., Bohr, V.A., Stevnsner, T., 2012. Mitochondrial base excision repair in mouse synaptosomes during normal aging and in a model of Alzheimer's disease. Neurobiol. Aging 33, 694–707.

Haleagrahara, N., Siew, C.J., Mitra, N.K., Kumari, M., 2011. Neuroprotective effect of bioflavonoid quercetin in 6-hydroxydopamine-induced oxidative stress biomarkers in the rat striatum. Neurosci. Lett. 500, 139–143.

Heeb, S., Fletcher, M.P., Chhabra, S.R., Diggle, S.P., Williams, P., Camara, M., 2011. Quinolones: from antibiotics to autoinducers. FEMS Microbiol. Rev. 35, 247–274.

Hu, Q., Uversky, V.N., Huang, M., Kang, H., Xu, F., Liu, X., et al., 2016. Baicalein inhibits α-synuclein oligomer formation and prevents progression of α-synuclein accumulation in a rotenone mouse model of Parkinson's disease. Biochim. Biophys. Acta (BBA) – Mol. Basis Dis. 1862, 1883–1890.

Huang, J.Z., Chen, Y.Z., Su, M., Zheng, H.F., Yang, Y.P., Chen, J., et al., 2010. DL-3-n-butylphthalide prevents oxidative damage and reduces mitochondrial dysfunction in an MPPC-induced cellular model of Parkinson's disease. Neurosci. Lett. 475, 89–94.

Javed, H., Azimullah, S., AbulKhair, S.B., Ojha, S., Haque, M.E., 2016. Neuroprotective effect of nerolidol against neuroinflammation and oxidative stress induced by rotenone. BMC Neurosci. 17, 58.

Jenner, P., 2003. Oxidative stress in Parkinson's disease. Ann. Neurol. 53, 26–38.

Jeong, J.K., Moon, M.H., Bae, B.C., Lee, Y.J., Seol, J.W., Kang, H.S., et al., 2012. Autophagy induced by resveratrol prevents human prion protein-mediated neurotoxicity. Neurosci. Res. 73, 99–105.

Kavitha, M., Manivasagam, T., Essa, M.M., Tamilselvam, K., Selvakumar, G.P., Karthikeyan, S., et al., 2014. Mangiferin antagonizes rotenone: induced apoptosis through attenuating mitochondrial dysfunction and oxidative stress in SK-N-SH neuroblastoma cells. Neurochem. Res. 39, 668–676.

Kim, M.E., Jung, I., Lee, J.S., Na, J.Y., Kim, W.J., Kim, Y.O., et al., 2017. Pseudane-vii isolated from *Pseudoalteromonas* sp. M2 ameliorates LPS-induced inflammatory response *in vitro* and *in vivo*. Mar. Drugs 15.

Kim, M.E., Jung, I., Na, J.Y., Lee, Y., Lee, J., Lee, J.S., et al., 2018. Pseudane-vii regulates LPS-induced neuroinflammation in brain microglia cells through the inhibition of iNOS expression. Molecules 23.

Lagouge, M., Argmann, C., Gerhart-Hines, Z., Meziane, H., Lerin, C., Daussin, F., et al., 2006. Resveratrol improves mitochondrial function and protects against metabolic disease by activating SIRT1 and PGC-1a. Cell 127, 1109–1122.

Lee, E., Park, H.R., Ji, S.T., Lee, Y., Lee, J., 2014. Baicalein attenuates astroglial activation in the 1-methyl-4-phenyl-1,2,3,4-tetrahydropyridine-induced Parkinson's disease model by downregulating the activations of nuclear factor-κB, ERK, and JNK. J. Neurosci. Res. 92, 130–139.

Lee, Y., Park, H.R., Chun, H.J., Lee, J., 2015. Silibinin prevents dopaminergic neuronal loss in a mouse model of Parkinson's disease via mitochondrial stabilization. J. Neurosci. Res. 93 (5), 755–765.

Li, L., Zhang, B., Tao, Y., Wang, Y., Wei, H., Zhao, J., et al., 2009. DL-3-nbutylphthalide protects endothelial cells against oxidative/nitrosative stress, mitochondrial damage and subsequent cell death after oxygen glucose deprivation *in vitro*. Brain Res. 1290, 91–101.

Li, S., Pu, X.P., 2011. Neuroprotective effect of kaempferol against a 1-methyl-4-phenyl-1,2,3,6- tetrahydropyridine-induced mouse model of Parkinson's disease. Biol. Pharm. Bull. 34, 1291–1296.

Lim, R., Morwood, C.J., Barker, G., Lappas, M., 2014. Effect of silibinin in reducing inflammatory pathways in in vitro and in vivo models of infection-induced preterm birth. PLoS One 9 (3), e92505.

Liu, C.M., Ma, J.Q., Liu, S.S., Zheng, G.H., Feng, Z.J., Sun, J.M., 2014. Proanthocyanidins improves lead-induced cognitive impairments by blocking endoplasmic reticulum stress and nuclear factor-κB-mediated inflammatory pathways in rats. Food Chem. Toxicol. 72, 295–302.

Lou, H., Jing, X., Wei, X., Shi, H., Ren, D., Zhang, X., 2014. Naringenin protects against 6-OHDA-induced neurotoxicity via activation of the Nrf2/ARE signaling pathway. Neuropharmacology 79, 380–3882.

Li, F.Q., Wang, T., Pei, Z., Liu, B., Hong, J.S., 2005. Inhibition of microglial activation by the herbal flavonoid baicalein attenuates inflammation-mediated degeneration of dopaminergic neurons. J. Neural Transm. 112, 331–347.

Lotharius, J., O'Malley, K.L., 2000. The Parkinsonism-inducing drug 1- Methyl-4-phenylpyridinium triggers intracellular dopamine oxidation. J. Biol. Chem. 275, 38581–38588.

Lu, X.L., Lin, Y.H., Wu, Q., Su, F.J., Ye, C.H., Shi, L., et al., 2015. Paeonolum protects against MPPC-induced neurotoxicity in zebrafish and PC12 cells. BMC Complement Altern. Med. 15, 137.

Lull, M.E., Block, M.L., 2010. Microglial activation and chronic neurodegeneration. Neurotherapeutics 7, 354–365.

Makino, T., Hishida, A., Goda, Y., Mizukami, H., 2008. Comparison of the major flavonoid content of *S. Baicalensis*, *S. Lateriflora*, and their commercial products. J. Nat. Med. 62, 294–299.

Mamelak, M., 2018. Parkinson's disease, the dopaminergic neuron and gammahydroxybutyrate. Neurol. Ther. 7, 5–11.

Manchope, M.F., Calixto-Campos, C., Coelho-Silva, L., Zarpelon, A.C., Pinho-Ribeiro, F.A., Georgetti, S.R., et al., 2016. Naringenin inhibits superoxide anion-induced inflammatory pain: role of oxidative stress, cytokines, Nrf-2 and the NO–cGMP–PKG–KATP channel signaling pathway. PLoS One 11, e0153015.

Meister, A., Anderson, M.E., 1983. Glutathione. Annu. Rev. Biochem. 52, 711–760.

Miean, K.H., Mohamed, S., 2001. Flavonoid (myricetin, quercetin, kaempferol, luteolin, and apigenin) content of edible tropical plants. J. Agric. Food Chem. 49, 3106–3112.

Moore, A.H., Bigbee, M.J., Boynton, G.E., Wakeham, C.M., Rosenheim, H.M., Staral, C.J., et al., 2010. Non-steroidal anti-inflammatory drugs in Alzheimer's disease and Parkinson's disease: reconsidering the role of neuroinflammation. Pharmaceuticals 3, 1812–1841.

Nabavi, S.F., Braidy, N., Habtemariam, S., Orhan, I.E., Daglia, M., Manayi, A., et al., 2015. Neuroprotective effects of chrysin: from chemistry to medicine. Neurochem. Int. 90, 224–231.

Ng, C.H., Guan, M.S., Koh, C., Ouyang, X., Yu, F., Tan, E.K., et al., 2012. AMP kinase activation mitigates dopaminergic dysfunctionand mitochondrial abnormalities in drosophila models of Parkinson'sdisease. J. Neurosci. 32, 14311–14317.

Obrenovich, M.E., Nair, N.G., Beyaz, A., Aliev, G., Reddy, V.P., 2010. The role of polyphenolic antioxidants in health, disease, and aging. Rejuvenation Res. 13, 631–643.

Obulesu, M., Jhansilakshmi, 2014. Neuroinflammation in Alzheimer's disease: an understanding of physiology and pathology. Int. J. Neurosci. 124, 227–235.

Obulesu, M., Venu, R., Somashekhar, R., 2011. Lipid peroxidation in Alzheimer's disease: emphasis on metal mediated neurotoxicity. Acta Neurol. Scand. 124, 295–301.

Pearce, R.K., Owen, A., Daniel, S., Jenner, P., Marsden, C.D., 1997. Alterations in the distribution of glutathione in the substantia nigra in Parkinson's disease. J. Neural Transm. 104, 661–677.

Postuma, R.B., Lang, A.E., Munhoz, R.P., Charland, K., Pelletier, A., Moscovich, M., et al., 2012. Caffeine for treatment of Parkinson disease: a randomized controlled trial. Neurology 79, 651–658.

Postuma, R.B., Anang, J., Pelletier, A., Joseph, L., Moscovich, M., Grimes, D., et al., 2017. Caffeine as symptomatic treatment for Parkinson disease (Cafe-PD): a randomized trial. Neurology 89, 1795–1803.

Przedborski, S., 2017. The two-centry journey of Parkinson disease research. Nat. Rev. Neurosci. 18, 251–259.

Rahimmi, A., Khosrobakhsh, F., Izadpanah, E., Moloudi, M.R., Hassanzadeh, K., 2015. N-acetylcysteine prevents rotenone-induced Parkinson's disease in rat: an investigation into the interaction of parkin and Drp1 proteins. Brain Res. Bull. 113, 34–40.

Reznichenko, L., Amit, T., Zheng, H., Avramovich-Tirosh, Y., Youdim, M.B., Weinreb, O., et al., 2006. Reduction of iron-regulated amyloid precursor protein and b-amyloid peptide by (-)-epigallocatechin-3-gallate in cell cultures: implications for iron chelation in Alzheimer's disease. J. Neurochem. 97, 527–536.

Shadel, G.S., Horvath, T.L., 2015. Mitochondrial ROS signaling in organismal homeostasis. Cell 163, 560–569.

Singh, N.A., Mandal, A.K., Khan, Z.A., 2016. Potential neuroprotective properties of epigallocatechin-3-gallate (EGCG). Nutr. J. 15, 60.

Smeriglio, A., Barreca, D., Bellocco, E., Trombetta, D., 2016. Chemistry, pharmacology and health benefits of anthocyanins. Phytother Res. 30, 1265–1286.

Sohal, R.S., Weindruch, R., 1996. Oxidative stress, caloric restriction, and aging. Science 273, 59–63.

Song, J.X., Shaw, P.C., Sze, C.W., Tong, Y., Yao, X.S., Ng, T.B., et al., 2010. Chrysotoxine, a novel bibenzyl compound, inhibits 6-hydroxydopamine induced apoptosis in SH-SY5Y cells via mitochondria protection and NFkB modulation. Neurochem. Int. 57, 676–689.

Song, J.X., Sze, S.C., Ng, T.B., Lee, C.K., Leung, G.P., Shaw, P.C., et al., 2012. Anti-Parkinsonian drug discovery from herbal medicines: what have we got from neurotoxic models. J. Ethnopharmacol. 139, 698–711.

Song, X., Zhou, B., Cui, L., Lei, D., Zhang, P., Yao, G., et al., 2017. Silibinin ameliorates Aβ25-35-induced memory deficits in rats by modulating autophagyand attenuating neuroinflammation as well as oxidative stress. Neurochemical Research 42 (4), 1073–1083.

Soobrattee, M.A., Bahorun, T., Aruoma, O.I., 2010. Chemopreventive actions of polyphenolic compounds in cancer. Biofactors 27, 19–35.

Stokes, A.H., Hastings, T.G., Vrana, K.E., 1999. Cytotoxic and genotoxic potential of dopamine. J. Neurosci. Res. 55, 659–665.

Svagera, Z., Skottova, N., Vana, P., Vecera, R., Urbanek, K., Belejova, M., et al., 2003. Plasma lipoproteins in transport of silibinin, an antioxidant flavonolignan from *Silybum marianum*. Phytother Res. 17, 524–530.

Wang, P., Niu, L., Gao, L., Li, W.X., Jia, D., Wang, X.L., et al., 2010. Neuroprotective effect of gypenosides against oxidative injury in the substantia nigra of a mouse model of Parkinson's disease. J. Int. Med. Res. 38, 1084–1092.

Wang, J., He, C., Wu, W.Y., Chen, F., Wu, Y.Y., Li, W.Z., et al., 2015. Biochanin A protects dopaminergic neurons against lipopolysaccharide-induced damage and oxidative stress in a rat model of Parkinson's disease. Pharmacol. Biochem. Behav. 138, 96–103.

Wang, M., Li, Y.J., Ding, Y., Zhang, H.N., Sun, T., Zhang, K., et al., 2016. Silibinin prevents autophagic cell death upon oxidative stress in cortical neurons and cerebral ischemia-reperfusion injury. Mol. Neurobiol. 53, 932–943.

Wang, S.Y., Chen, C.T., Sciarappa, W., Wang, C.Y., Camp, M.J., 2008. Fruit quality, antioxidant capacity, and flavonoid content of organically and conventionally grown blueberries. J. Agric. Food Chem. 56, 5788–5794.
Winter, A.N., Ross, E.K., Wilkins, H.M., Stankiewicz, T.R., Wallace, T., Miller, K., et al., 2017. An anthocyanin-enriched extract from strawberries delays disease onset and extends survival in the hSOD1G93A mouse model of amyotrophic lateral sclerosis. Nutr. Neurosci. 9, 1–13.
Ward, R.J., Zucca, F.A., Duyn, J.H., Crichton, R.R., Zecca, L., 2014. The role of iron in brain ageing and neurodegenerative disorders. Lancet Neurol. 13, 1045–1060.
Weinreb, O., Amit, T., Mandel, S., Youdim, M.B., 2009. Neuroprotective molecular mechanisms of (2)-epigallocatechin- 3-gallate: a reflective outcome of its antioxidant, iron chelating and neuritogenic properties. Genes Nutr. 4, 283–296.
Xu, H., Jiang, H., Wang, J., Xie, J., 2010. Rg1 protects iron-induced neurotoxicity through antioxidant and iron regulatory proteins in 6-OHDA treated MES23.5 cells. J. Cell. Biochem. 111, 1537–1545.
Ye, Q., Ye, L., Xu, X., Huang, B., Zhang, X., Zhu, Y., et al., 2012. Epigallocatechin-3-gallate suppresses 1-methyl-4-phenyl-pyridine-induced oxidative stress in PC12 cells via the SIRT1/PGC-1alpha signaling pathway. BMC Complement Altern. Med. 12, 82.
Zhang, J., Liu, Q., Chen, Q., Liu, N.Q., Li, F.L., Lu, Z.B., et al., 2006. Nicotine attenuates beta-amyloid-induced neurotoxicity by regulating metal homeostasis. FASEB J. 20, 1212–1214.
Zhang, X., Xu, J.K., Wang, J., Wang, N.L., Kurihara, H., Kitanaka, S., et al., 2007. Bioactive bibenzyl derivatives and fluorenones from dendrobium nobile. J. Nat. Prod. 70, 24–28.
Zuk, M., Kulma, A., Dyminska, L., Szołtysek, K., Prescha, A., Hanuza, J., et al., 2011. Flavonoid engineering of flax potentiate its biotechnological application. BMC Biotechnol. 12, 47.

CHAPTER 5

Curcumin: a promising therapeutic in Parkinson's disease treatment

1. Introduction

Curcumin or diferuloylmethane is the essential compound isolated from *Curcuma longa* rhizome (Jiang et al., 2013; Yu et al., 2016; Sang et al., 2018). Interestingly, it offers multifarious therapeutic benefits such as antioxidant, antiinflammatory, antitumor, and antinerve disintegration (Yang et al., 2014; van der Merwe et al., 2017; Sang et al., 2018). Curcumin confers protection against neurodegeneration with constant and adequate intake (Gota et al., 2010, Mythri and Srinivas-Bharath, 2012, Fu et al., 2015, Zhang et al., 2018). Additionally, it is not toxic at higher concentrations also (Pan et al., 2012; Qualls et al., 2014; Storka et al., 2015; Zhang et al., 2018).

Mounting evidence has also shown that curcumin curtails glial fibrillary acidic protein level in Parkinson's disease (PD) animal models compared to control group (Pan et al., 2007, Yu et al., 2012, Tripanichkul and Jaroensuppaperch, 2013, Wang et al., 2017a). Additionally, it also ameliorated reduced libido (sex drive), failure to orgasm, erectile dysfunction, and untimely ejaculation, which are usual sex problems in PD (Wu et al., 2017, Akintunde et al., 2018).

2. Antioxidant activity

It also ameliorated hypothalamus—pituitary—testicular hormones by regulating acetylcholinesterase activity, locomotion, intracellular nitric oxide (NO) level decrease and averting striatum-endocrine injury and oxidative damage in bisphenol A—induced oxidative damage (Akintunde et al., 2018). It exhibited neuroprotection in numerous cell and animal PD

Parkinson's Disease Therapeutics
ISBN 978-0-12-819882-7
https://doi.org/10.1016/B978-0-12-819882-7.00005-2

models by controlling free radical production and reactive oxygen species (ROS) such as malondialdehyde (Chen et al., 2006; Jagatha et al., 2008; Mythri et al., 2011; Mythri and Srinivas-Bharath, 2012, Akintunde et al., 2018). Interestingly, chronic dietary intake of curcumin was also observed to instigate neuroprotection in a PD mouse model (Mythri and Srinivas-Bharath, 2012, Akintunde et al., 2018). Curcumin also combated oxidative stress—mediated injury in PD rats by initiating wingless-type mouse mammary tumor virus integration site (Wnt)/β-catenin signaling pathway, which has been found to be responsible for PD pathogenesis (L'Episcopo et al., 2014, Wang et al., 2017b).

3. 6-Hydroxydopamine and MPTP PD models

Curcumin showed remarkable protection by scavenging ROS in 6-hydroxydopamine (6-OHDA) and 1-methyl-4-phenyl-1,2,3,6-tetrahydropyridine (MPTP)-induced PD models (Zbarsky et al., 2005; Yang et al., 2014, Akintunde et al., 2018). Its treatment has been found to recover mitochondrial membrane potential and enhance Cu/Zn superoxide dismutase (SOD) and cell viability in 6-OHDA-lesioned mouse embryonic stem cells (Wang et al., 2009; Wang et al., 2017a). In line with this, Rajeswari et al. (2008) have reported that striatal dopamine and 3,4-dihydroxyphenylacetic acid levels were enhanced after curcumin injection in MPTP-treated mice (Wang et al., 2017a). Guo et al. (2012) reported the significant protective efficacy of curcumin against 6-OHDA-induced SOD1 expression reduction (Wang et al., 2017a).

4. α-Synuclein targeting

Wealth of studies has shown the neuroprotective efficacy of curcumin in PC12 cells treated with A53T α-synuclein and in vivo also (Liu et al., 2011; Cui et al., 2016; Sang et al., 2018). Accordingly, Cui et al. (2016) showed the reduction of dopaminergic neuron loss by augmenting tyrosine hydroxylase activity. It also decreased leucine-rich repeat kinase 2 (LRRK2) activity in LRRK2-initiated PD cell models and facilitated the protection of LRRK2 transgenic *Drosophila* (Yang et al., 2012; Sang et al., 2018). In another study, Singh et al. (2013) reported that curcumin attached to preformed oligomers and fibrils averted aggregation of α-synuclein (Sang et al., 2018). It protected against PD by affecting the stability of α-synuclein protein (Liu et al., 2014, Akintunde et al., 2018).

5. Curcumin derivatives

Demethoxycurcumin (DMC), a derivative of curcumin, also ameliorated motor dysfunctions, neurochemical deficits, and oxidative stress in rotenone-treated rats (Ramkumar et al., 2018). In addition, DMC and bisdemethoxycurcumin (BDMC) have been used in food industry for cooking and folk medicine in combination with curcumin (Guo et al., 2008; Ramkumar et al., 2018). Interestingly, these derivatives have also been found to have profound antioxidant, antiproliferative, and antiinflammatory activities similar to curcumin (Simon et al., 1998; Kim et al., 2001; Sandur et al., 2007; Ramkumar et al., 2018). DMC has been found to have higher antiinflammatory and anticancer activities compared with curcumin (Zhang et al., 2008; Lee et al., 2010; Ramkumar et al., 2018). DMC also ameliorates the activities of catalase, SOD, and glutathione peroxidase in 6-OHDA-induced rat model of PD (Agrawal et al., 2012; Ramkumar et al., 2018). It also showed substantial suppression of NO and TNF-α generation compared to curcumin in lipopolysaccharide-induced rat primary microglia (Zhang et al., 2010; Ramkumar et al., 2018). Curcumin and its metabolite tetrahydrocurcumin (ThC) showed neuroprotection against MPTP-initiated neurotoxicity by attenuating monoamine oxidase B activity (Rajeswari and Sabesan, 2008; Ding et al., 2018).

6. Bioavailability

A few studies have shown the low bioavailability of curcumin due to insufficient absorption and unstable metabolic activity (Darvesh et al., 2012; Ahsan et al., 2015; Sang et al., 2018). To overcome these issues, a few stable curcumin analogs such as curcumin pyrazole and its derivative N-(3-nitrophenylpyrazole) curcumin were used to mitigate neurotoxicity in neurodegenerative diseases such as PD (Sang et al., 2018). Studies have also shown that curcuminoids such as curcumin, DMC, and BDMC ameliorated dopaminergic neurodegeneration in MPTP-initiated PD models (Ojha et al., 2012; Sang et al., 2018). Along with curcumin, semisynthetic curcuminoids and associated metal complexes were also found to be substantial PD therapeutic agents (Sang et al., 2018).

7. Heat shock proteins

Among the heat shock proteins (HSPs), 90-kda HSPs (HSP90) act as molecular chaperons which play a pivotal role in protein folding, stabilization, and initiation (Daturpalli et al., 2013; Alani et al., 2014; Sang et al., 2018). While the HSPs are at low levels in healthy conditions, they are increased in cellular stress (Aridon et al., 2011; Sang et al., 2018). HSP90, a crucial therapeutic target in cancer, has been recently focused for its therapeutic role in neurodegenerative diseases also (Daturpalli et al., 2013; Sang et al., 2018). In line with this, Aridon et al., (2011) reported that HSP70 and HSP90 curbed protein misfolding and suppressed neuronal apoptosis (Aridon et al., 2011; Sang et al., 2018). HSP60 was found to ameliorate PD pathogenesis in 6-OHDA-lesioned rat models (Zhao et al., 2016; Sang et al., 2018). Curcumin showed substantial protective efficacy by attenuating the toxic effects of 1-methyl-4-phenyl-pyridinium (MPP^+) on SH-SY5Y cells and remarkably decreasing effects of MPP^+ on dopaminergic neurons by upregulating HSP90 (Sang et al., 2018).

8. Antiapoptosis

Several lines of evidence have shown that curcumin downregulated p53 in 6-OHDA-treated SH-SY5Y cell lines through inhibition of p53 phosphorylation and fulfillment of balance between anti- and proapoptotic Bcl2-family proteins (Jaisin et al., 2011; Wang et al., 2017a). It has been proven that curcumin increases striatal dopamine levels and averts neuronal apoptosis (Wang et al., 2017a). It has also been found to intervene with the Bcl2 signal pathway and mitigate the proapoptotic proteins such as Bax and Bad in PC12 cells treated with MPP^+ (Wang et al., 2017a). It also showed neuroprotective effects on the MPTP-initiated cellular model by regulating Bcl2-mitochondria–ROS–iNOS pathway (Chen et al., 2006; Ding et al., 2018). It attenuates astrocyte activation of NADPH oxidase complex initiation by inhibiting NF-κB activity, intrinsic apoptotic pathway (Bax, Bcl-2, caspase 3, and caspase 9), proinflammatory cytokines (TNF-α, IL-1β, and IL-1α), and inducible nitric oxide synthase (iNOS) activity (Sharma and Nehru, 2017, Akintunde et al., 2018).

9. Metal chelation

Pretreatment with curcumin ameliorated iron-provoked disintegration of nigral dopaminergic neurons through its iron chelating efficacy (Du et al., 2012; Ding et al., 2018). Additionally, oxidative DNA damage and profound attenuation of DNA repair enzymes by metal ions such as copper (Cu) and iron (Fe) were observed in PD (Grin et al., 2009; Wang et al.,

2017a). Nevertheless, curcumin was found to protect DNA repair enzymes significantly (Wang et al., 2017a).

10. Delivery systems for curcumin

Nanotechnology has been extensively used to improve blood–brain barrier (BBB) spanning, solubility, and stability of bioactive molecules and target the same to the brain (Tsai et al., 2011a,b, Podsedek et al., 2014; Ganesan et al., 2015). The vital nanoparticle (NP) antioxidant mechanism includes shortening of natural bioactive compound (curcumin, vitamin E, or resveratrol) to a nanosize that enhances absorption and targeting, in turn restoring the activity (Cheng et al., 2013; Yao et al., 2014; Coradini et al., 2014; Ganesan et al., 2015). Shortening of curcumin and conjugating it with polyesters resulted in enhanced bioavailability in systemic circulation (Thangapazham et al., 2008; Takahashi et al., 2009; Mourtas et al., 2014; Young et al., 2014; Ganesan et al., 2015). The key characteristics of NPs targeted to the brain are bioactive compound size, carrier toxicity, and surface activity (Fathi et al., 2012; Frozza et al., 2013; Kumar et al., 2013; Ganesan et al., 2015).

To overcome low bioavailability and inability of curcumin to cross BBB multifarious, drug delivery systems (DDS) have been studied. Voluminous data have shown that design of nanosized curcumin particles or its dissolution in dimethyl sulfoxide profoundly enhances its solubility in vivo (Wang et al., 2017a). Multifarious delivery systems employed to enhance the bioavailability of curcumin include NPs, liposomes, and micelles (Ji and Shen, 2014; Wang et al., 2017a). Curcumin-encapsulated polysorbate 80 (PS80) modified cerasome NPs in association with ultrasound-targeted microbubble destruction (CPC-UTMD) showed increased curcumin levels in the brain and understood as viable therapeutic option for PD (Zhang et al., 2018). In this formulation, hydrophobic curcumin was localized in the hydrophobic moiety of the cerasomal lipid bilayer (Ma et al., 2011; Liang et al., 2011, 2015; Blanco et al., 2015; Zhang et al., 2018). Surface functionalization with PS80 was used to facilitate the curcumin nanocarriers span the BBB by transcytosis (Wagner et al., 2012; Zhang et al., 2018). Additionally, bilayer vesicular structure and silicate surface of cerasome is similar to lipids but confers more stability than liposomes and substantial biocompatibility than silica NPs (Zhang et al., 2018). Therefore, based on substantial stability to PS80, the liposomal nanohybrid cerasome has been considered the potential curcumin nanocarrier (Liang et al., 2013; Cao et al., 2012, 2013; Yue and Dai, 2014; Zhang et al., 2018). Curcumin-encapsulated nanocapsules inhibited oxidation or hydroxylation of curcumin in the body (Tsai et al., 2011a; Ganesan et al., 2015) (Table 5.1). Oral administration of curcumin-loaded nanoliposomes conferred enhanced bioavailability and increased antioxidant activity in rats (Takahashi et al., 2009; Ganesan et al., 2015). In an in vitro study, curcumin nanoliposomes showed significant BBB penetration and therapeutic effect against AD (Mourtas et al., 2014; Ganesan et al., 2015).

In a transgenic *Drosophila* PD model, an alginate curcumin nanocomposite showed neuroprotective effect by decreasing oxidative stress and apoptosis (Siddique et al., 2014; Ganesan et al., 2015). Among the numerous nanodelivery systems, curcumin-encapsulated polylactic-co-glycolic acid (PLGA) NPs offered increased bioavailability (Ganesan et al., 2015). Tsai et al. (2011a,b) emphasized that nanocurcumin is bioavailable in the blood and plasma and can ferry BBB feasibly (Ganesan et al., 2015). The bioavailability of solid lipid nanocurcumin is profoundly increased in the mouse brain and rendered remarkable pharmacological activity (Ramalingam and Ko, 2014, 2015; Ganesan et al., 2015). Accordingly, increased bioavailability of nanocurcumin in mouse brain was found to combat oxidative stress (Nazari et al., 2014; Ganesan et al., 2015).

11. Conclusions and future perspectives

Curcumin confers appreciable antioxidant, antiinflammatory, and antitumor activities. Despite its remarkable therapeutic efficacy, low bioavailability, solubility, and short systemic circulation limited its success. Although curcumin showed appreciable anti-PD properties, yet the underlying molecular mechanisms are elusive (Sang et al., 2018). In addition, randomized controlled clinical trials are essential to confirm its robust therapeutic efficacy (Wang et al., 2017a). Several DDS used to overcome the bioavailability issues include PLGA NPs, nanoliposomes, nanocomposites, and nanocurcumin. They have remarkably improved the bioavailability of curcumin. Among multifarious DDS employed till date, PLGA NPs showed better therapeutic efficacy compared to others. However, garnering the expertise from multifarious fields such as medicine, chemistry, and nanotechnology, there is a pressing need to develop more appropriate delivery systems.

Table 5.1 List of curcumin, curcumin derivatives, and curcumin delivery systems and their therapeutic effects.

Therapeutic compound	PD model	Therapeutic effect	Reference
Curcumin	PD mouse model	Reduced glial fibrillary acidic protein levels	Pan et al. (2007), Yu et al. (2012), Tripanichkul and Jaroensuppaperch, 2013, Wang et al. (2017a)
	PD Patients	Protection against reduced libido (sex drive), failure to orgasm, erectile dysfunction, untimely ejaculation	Akintunde et al., 2018
	PD Cell and Animal models	Controlling free radical production and ROS such as malondialdehyde	Chen et al. (2006), Jagatha et al. (2008), Mythri et al. (2011), Mythri and Srinivas-Bharath (2012), Akintunde et al., 2019
	PD rats	Combated oxidative-stress mediated injury by initiating wingless-type mouse mammary tumor virus integration site (Wnt)/β-catenin signaling pathway	L'Episcopo et al. (2014), Wang et al., 2017b
	6-OHDA−MES cells	Recover mitochondrial membrane potential (MMP), enhance Cu/Zn superoxide dismutase (SOD) and cell viability	Wang et al. (2009), Wang et al. (2017a)
	MPTP treated mice	Enhanced striatal dopamine and 3,4-dihydroxyphenylacetic acid levels	Wang et al. (2017a)
	6-OHDA-treated SH-SY5Y cell lines	Downregulated p53, inhibited p53 phosphorylation and balanced anti and proapoptotic Bcl2 family proteins	Jaisin et al. (2011), Wang et al. (2017a)
	Nigral DA neurons	Ameliorated iron-provoked disintegration of nigral DA neurons	Du et al. (2012), Ding et al. (2018)

Continued

Table 5.1 List of curcumin, curcumin derivatives, and curcumin delivery systems and their therapeutic effects.—cont'd

Therapeutic compound	PD model	Therapeutic effect	Reference
Curcumin derivatives			
Demethoxycurcumin	Rotenone-treated rats	Ameliorated motor dysfunctions, neurochemical deficits and oxidative stress	Ramkumar et al. (2018)
	6-OHDA-induced rat model of PD	Amelioration of catalase, SOD, and glutathione peroxidase	Agrawal et al. (2012), Ramkumar et al. (2018)
	Lipopolysaccharide-induced rat primary microglia	Suppression of NO and TNF-α generation	Zhang et al. (2010), Ramkumar et al. (2018)
Drug delivery systems for curcumin			
CPC-UTMD	PD mouse model	Increased brain curcumin levels and enhanced therapeutic efficacy	Zhang et al. (2018)
Curcumin nanocapsules	Rats	Inhibited oxidation or hydroxylation of curcumin in the body	Tsai et al. (2011a), Ganesan et al. (2015)
Curcumin-loaded nanoliposomes	Rats	Conferred enhanced bioavailability, increased antioxidant activity	Takahashi et al. (2009), Ganesan et al. (2015)
Curcumin nanocomposite	Transgenic *Drosophila* PD model	Decreased oxidative stress and apoptosis	Siddique et al. (2014), Ganesan et al. (2015)
PLGA NPs	In vitro and in vivo	Increased bioavailability	Ganesan et al. (2015)
Nanocurcumin	Mouse model	Increased bioavailability	Ramalingam and Ko (2014), 2015, Ganesan et al. (2015)

References

Agrawal, S.S., Gullaiya, S., Dubey, V., Singh, V., Kumar, A., Nagar, A., et al., 2012. Neurodegenerative shielding by curcumin and its derivatives on brain lesions induced by 6-OHDA model of Parkinson's disease in albino wistar rats. Cardiovasc. Psychiatry Neurol. 942981.

Ahsan, N., Mishra, S., Jain, M.K., Surolia, A., Gupta, S., 2015. Curcumin Pyrazole and its derivative (N-(3-Nitrophenylpyrazole) Curcumin inhibit aggregation, disrupt fibrils and modulate toxicity of wild type and mutant alpha-Synuclein. Sci. Rep. 5, 9862.

Akintunde, J.K., Farouk, A.A., Mogbojuri, O., 2018. Metabolic treatment of syndrome linked with Parkinson's disease and hypothalamus pituitarygonadal hormones by turmeric curcumin in Bisphenol-A induced neuro-testicular dysfunction of wistar rat. Biochem. Biophys. Rep. 17, 97–107.

Alani, B., Salehi, R., Sadeghi, P., Zare, M., Khodagholi, F., Arefian, E., et al., 2014. Silencing of Hsp90 chaperone expression protects against 6-hydroxydopamine toxicity in PC12 cells. J. Mol. Neurosci. 52, 392–402.

Aridon, P., Geraci, F., Turturici, G., D'Amelio, M., Savettieri, G., Sconzo, G., 2011. Protective role of heat shock proteins in Parkinson's disease. Neurodegener. Dis. 8, 155–168.

Blanco, E., Shen, H., Ferrari, M., 2015. Principles of nanoparticle design for overcoming biological barriers to drug delivery. Nat. Biotechnol. 33, 941–951.

Cao, Z., Yue, X., Jin, Y., Wu, X., Dai, Z., 2012. Modulation of release of paclitaxel from composite cerasomes. Colloids Surf. B 98, 97–104.

Cao, Z., Yue, X., Li, X., Dai, Z., 2013. Stabilized magnetic cerasomes for drug delivery. Langmuir 29, 14976–14983.

Chen, J., Tang, X.Q., Zhi, J.L., Cui, Y., Yu, H.M., Tang, E.H., et al., 2006. Curcumin protects PC12 cells against 1-methyl-4-phenylpyridinium ion-induced apoptosis by bcl-2-mitochondria-ROS-iNOS pathway. Apoptosis 11, 943–953.

Cheng, K.K., Yeung, C.F., Ho, S.W., Chow, S.F., Chow, A.H.L., Baum, L., 2013. Highly stabilized curcumin nanoparticles tested in an in vitro blood-brain barrier model and in Alzheimer's disease Tg2576 mice. AAPS J. 15, 324–336.

Coradini, K., Lima, F.O., Oliveira, C.M., Chaves, P.S., Athayde, M.L., Carvalho, L.M., et al., 2014. Co-encapsulation of resveratrol and curcumin in lipid-core nanocapsules improves their in vitro antioxidant effects. Eur. J. Pharm. Biopharm. 88, 178–185.

Cui, Q., Li, X., Zhu, H., 2016. Curcumin ameliorates dopaminergic neuronal oxidative damage via activation of the Akt/Nrf2 pathway. Mol. Med. Rep. 13, 1381–1388.

Darvesh, A.S., Carroll, R.T., Bishayee, A., Novotny, N.A., Geldenhuys, W.J., Van der Schyf, C.J., 2012. Curcumin and neurodegenerative diseases: a perspective. Expert Opin. Investig. Drugs 21, 1123–1140.

Daturpalli, S., Waudby, C.A., Meehan, S., Jackson, S.E., 2013. Hsp90 inhibits alpha-synuclein aggregation by interacting with soluble oligomers. J. Mol. Biol. 425, 4614–4628.

Ding, Y., Xin, C., Zhang, C.W., Lim, K.L., Zhang, H., Fu, Z., et al., 2018. Natural molecules from Chinese herbs protecting against Parkinson's disease viaAnti-oxidativeStress. Front. Aging Neurosci. 10, 246.

Du, X.X., Xu, H.M., Jiang, H., Song, N., Wang, J., Xie, J.X., 2012. Curcumin protects nigral dopaminergic neurons by iron-chelation in the 6-hydroxydopamine rat model of Parkinson's disease. Neurosci. Bull. 28, 253–258.

Fathi, M., Mozafari, M.R., Mohebbi, M., 2012. Nanoencapsulation of food ingredients using lipid based delivery systems. Trends Food Sci. Technol. 23, 13–27.

Frozza, R.L., Bernardi, A., Hoppe, J.B., Meneghetti, A.B., Battastini, A.M., Pohlmann, A.R., et al., 2013. Lipid-core nanocapsules improve the effects of resveratrol against A beta-induced neuroinflammation. J. Biomed. Nanotechnol. 9, 2086–2104.

Fu, W., Zhuang, W., Zhou, S., Wang, X., 2015. Plant-derived neuroprotective agents in Parkinson's disease. Am. J. Transl. Res. 7, 1189–1202.

Ganesan, P., Ko, H.M., Kim, I.S., Choi, D.K., 2015. Recent trends in the development of nanophytobioactive compounds and delivery systems for their possible role in reducing oxidative stress in Parkinson's disease models. Int. J. Nanomed. 10, 6757–6772.

Gota, V.S., Maru, G.B., Soni, T.G., Gandhi, T.R., Kochar, N., Agarwal, M.G., 2010. Safety and pharmacokinetics of a solid lipid curcumin particle formulation in osteosarcoma patients and healthy volunteers. J. Agric. Food Chem. 58, 2095–2099.

Grin, I.R., Konorovsky, P.G., Nevinsky, G.A., Zharkov, D.O., 2009. Heavy metal ions affect the activity of DNA glycosylases of the fpg family. Biochemistry (Mosc.) 74, 1253–1259.

Guo, L.Y., Cai, X.F., Lee, J.J., Kang, S.S., Shin, E.M., Zhou, H.Y., et al., 2008. Comparison of suppressive effects of demethoxycurcumin and bisdemethoxycurcumin on expressions of inflammatory mediators *in vitro* and *in vivo*. Arch. Pharm. Res. 31, 490–496.

Guo, Y.X., Yang, B., Shi, L., Gu, J., Chen, H., 2012. Anti-inflammation mechanism of curcumin in mice with lipopolysaccharide-induced Parkinson's disease. J. Med. Post Grad. 25, 582–587.

Jagatha, B., Mythri, R.B., Vali, S., Bharath, M.M., 2008. Curcumin treatment alleviates the effects of glutathione depletion in vitro and in vivo: therapeutic implications for Parkinson's disease explained via in silico studies. Free Radic. Biol. Med. 44, 907–917.

Jaisin, Y., Thampithak, A., Meesarapee, B., Ratanachamnong, P., Suksamrarn, A., Phivthong-Ngam, L., et al., 2011. Curcumin I protects the dopaminergic cell line SH-SY5Y from 6-hydroxydopamine-induced neurotoxicity through attenuation of p53-mediated apoptosis. Neurosci. Lett. 489, 192–196.

Ji, H.F., Shen, L., 2014. Can improving bioavailability improve the bioactivity of curcumin? Trends Pharmacol. Sci. 35, 265–266.

Jiang, T.F., Zhang, Y.J., Zhou, H.Y., Wang, H.M., Tian, L.P., Liu, J., et al., 2013. Curcumin ameliorates the neurodegenerative pathology in A53T alpha-synuclein cell model of Parkinson's disease through the downregulation of mTOR/p70S6K signaling and the recovery of macroautophagy. J. Neuroimmune Pharmacol. 8, 356–369.

Kim, D.S., Park, S.Y., Kim, J.K., 2001. Curcuminoids from *Curcuma longa* L. (*Zingiberaceae*) that protect PC12 rat pheochromocytoma and normal human umbilical vein endothelial cells from betaA (1-42) insult. Neurosci. Lett. 303, 57–61.

Kumar, A., Chen, F., Mozhi, A., Zhang, X., Zhao, Y., Xue, X., et al., 2013. Innovative pharmaceutical development based on unique properties of nanoscale delivery formulation. Nanoscale 5, 8307–8325.

Lee, J.W., Hong, H.M., Kwon, D.D., Pae, H.O., Jeong, H.J., 2010. Dimethoxycurcumin, a structural analogue of curcumin, induces apoptosis in human renal carcinoma caki cells through the production of reactive oxygen species, the release of cytochrome C, and the activation of caspase-3. Korean J. Urol. 51, 870–878.

Liang, X., Li, X., Yue, X., Dai, Z., 2011. Conjugation of porphyrin to nanohybrid cerasomes for photodynamic therapy of cancer. Angew. Chem. Int. 50, 11622–11627.

Liang, X., Li, X., Jing, L., Xue, P., Jiang, L., Ren, Q., et al., 2013. Design and synthesis of lipidic organoalkoxysilanes for the self-assembly of liposomal nanohybrid cerasomes with controlled drug release properties. Chem. Euro. J. 19, 16113–16121.

Liang, X., Gao, J., Jiang, L., Luo, J., Jing, L., Li, X., et al., 2015. Nanohybrid liposomal cerasomes with good physiological stability and rapid temperature responsiveness for high intensity focused ultrasound triggered local chemotherapy of cancer. ACS Nano 9, 1280–1293.

Liu, Z., Yu, Y., Li, X., Ross, C.A., Smith, W.W., 2011. Curcumin protects against A53T alpha-synuclein-induced toxicity in a PC12 inducible cell model for Parkinsonism. Pharmacol. Res. 63, 439–444.

Liu, D., Wang, Z., Gao, Z., Xie, K., Zhang, Q., Jiang, H., et al., 2014. Effects of curcumin on learning and memory deficits, BDNF, and ERK protein expression in rats exposed to chronic unpredictable stress. Behav. Brain Res. 271, 116–121.

L'Episcopo, F., Tirolo, C., Caniglia, S., Testa, N., Morale, M.C., Serapide, M.F., et al., 2014. Targeting Wnt signaling at the neuroimmune interface for dopaminergic neuroprotection/repair in Parkinson's disease. J. Mol. Cell Biol. 6, 13–26.

Ma, Y., Dai, Z., Zha, Z., Gao, Y., Yue, X., 2011. Selective Antileukemia effect of stabilized nanohybrid vesicles based on cholesteryl succinyl silane. Biomaterials 32, 9300–9307.

Mourtas, S., Lazar, A.N., Markoutsa, E., Duyckaerts, C., Antimisiaris, S.G., 2014. Multifunctional nanoliposomes with curcumin-lipid derivative and brain targeting functionality with potential applications for Alzheimer disease. Eur. J. Med. Chem. 80, 175–183.

Mythri, R.B., Srinivas-Bharath, M.M., 2012. Curcumin: a potential neuroprotective agent in Parkinson's disease. Curr. Pharmacol. Des. 18, 91–99.

Mythri, R.B., Harish, G., Dubey, S.K., Misra, K., Bharath, M.M., 2011. Glutamoyl diester of the dietary polyphenol curcumin offers improved protection against peroxynitrite mediated nitrosative stress and damage of brain mitochondria in vitro: implications for Parkinson's disease. Mol. Cell. Biochem. 347, 135–143.

Nazari, Q.A., Takada-Takatori, Y., Hashimoto, T., Imaizumi, A., Izumi, Y., Akaike, A., et al., 2014. Potential protective effect of highly bioavailable curcumin on an oxidative stress model induced by microinjection of sodium nitroprusside in mice brain. Food Funct. 5, 984–989.

Ojha, R.P., Rastogi, M., Devi, B.P., Agrawal, A., Dubey, G.P., 2012. Neuroprotective effect of curcuminoids against inflammation-mediated dopaminergic neurodegeneration in the MPTP model of Parkinson's disease. J. Neuroimmune Pharmacol. 7, 609–618.

Pan, J., Ding, J., Chen, S., 2007. The protection of curcumin in nigral dopaminergic neuronal injury of mice model of Parkinson disease. Chin. J. Contemp. Neurol. Neurosurg. 7, 421–426.

Pan, J., Li, H., Ma, J.F., Tan, Y.Y., Xiao, Q., Ding, J.Q., et al., 2012. Curcumin inhibition of JNKs prevents dopaminergic neuronal loss in a mouse model of Parkinson's disease through suppressing mitochondria dysfunction. Transl. Neurodegener. 1, 16.

Podsedek, A., Redzynia, M., Klewicka, E., Koziolkiewicz, M., 2014. Matrix effects on the stability and antioxidant activity of red cabbage anthocyanins under simulated gastrointestinal digestion. BioMed Res. Int. 365738.

Qualls, Z., Brown, D., Ramlochansingh, C., Hurley, L.L., Tizabi, Y., 2014. Protective effects of curcumin against rotenone and salsolinol-induced toxicity: implications for Parkinson's disease. Neurotox. Res. 25, 81–89.

Rajeswari, A., Sabesan, M., 2008. Inhibition of monoamine oxidase-B by the polyphenolic compound, curcumin and its metabolite tetrahydrocurcumin in a model of Parkinson's disease induced by MPTP neurodegeneration in mice. Inflammopharmacology 16, 96–99.

Ramalingam, P., Ko, Y.T., 2014. A validated LC-MS/MS method for quantitative analysis of curcumin in mouse plasma and brain tissue and its application in pharmacokinetic and brain distribution studies. J. Chromatogr. B 969, 101–108.

Ramalingam, P., Ko, Y.T., 2015. Enhanced oral delivery of curcumin from N-trimethyl chitosan surface-modified solid lipid nanoparticles: pharmacokinetic and brain distribution evaluations. Pharm. Res. 32, 389–402.

Ramkumar, M., Rajasankar, S., Gobi, V.V., Janakiraman, U., Manivasagam, T., Thenmozhi, A.J., et al., 2018. Demethoxycurcumin, a natural derivative of curcumin abrogates rotenone-induced dopamine depletion and motor deficits by its antioxidative and anti-inflammatory properties in parkinsonian rats. Pharmacogn. Mag. 14, 9–16.

Sandur, S.K., Pandey, M.K., Sung, B., Ahn, K.S., Murakami, A., Sethi, G., et al., 2007. Curcumin, demethoxycurcumin, bisdemethoxycurcumin, tetrahydrocurcumin and turmerones differentially regulate anti-inflammatory and anti-proliferative responses through a ROS-independent mechanism. Carcinogenesis 28, 1765–1773.

Sang, Q., Liu, X., Wang, L., Qi, L., Sun, W., Wang, W., et al., 2018. Curcumin protects an SH-SY5Y cell model of Parkinson's disease against toxic injury by regulating HSP90. Cell. Physiol. Biochem. 51, 681–691.

Sharma, N., Nehru, B., 2017. Curcumin affords neuroprotection and inhibits α-synuclein aggregation in lipopolysaccharide induced Parkinson's disease model. Inflammopharmacology 26 (2), 349–360.

Siddique, Y.H., Naz, F., Jyoti, S., 2014. Effect of curcumin on lifespan, activity pattern, oxidative stress, and apoptosis in the brains of transgenic Drosophila model of Parkinson's disease. BioMed Res. Int. 606928.

Simon, A., Allais, D.P., Duroux, J.L., Basly, J.P., Durand-Fontanier, S., Delage, C., et al., 1998. Inhibitory effect of curcuminoids on MCF-7 cell proliferation and structure-activity relationships. Cancer Lett. 129, 111–116.

Singh, P.K., Kotia, V., Ghosh, D., Mohite, G.M., Kumar, A., Maji, S.K., 2013. Curcumin modulates alpha-synuclein aggregation and toxicity. ACS Chem. Neurosci. 4, 393–407.

Storka, A., Vcelar, B., Klickovic, U., Gouya, G., Weisshaar, S., Aschauer, S., et al., 2015. Safety, tolerability and pharmacokinetics of liposomal curcumin (Lipocurc™) in healthy humans. Int. J. Clin. Pharmacol. Therapeut. 53, 54–65.

Takahashi, M., Uechi, S., Takara, K., Asikin, Y., Wada, K., 2009. Evaluation of an oral carrier system in rats: bioavailability and antioxidant properties of liposome-encapsulated curcumin. J. Agric. Food Chem. 57, 9141–9146.

Thangapazham, R.L., Puri, A., Tele, S., Blumenthal, R., Maheshwari, R.K., 2008. Evaluation of a nanotechnology-based carrier for delivery of curcumin in prostate cancer cells. Int. J. Oncol. 32, 1119–1123.

Tripanichkul, W., Jaroensuppaperch, E.O., 2013. Ameliorating effects of curcumin on 6-OHDA-induced dopaminergic denervation, glial response, and SOD1 reduction in the striatum of hemiparkinsonian mice. Eur. Rev. Med. Pharmacol. Sci. 17, 1360–1368.

Tsai, Y.M., Jan, W.C., Chien, C.F., Lee, W.C., Lin, L.C., Tsai, T.H., 2011a. Optimised nano-formulation on the bioavailability of hydrophobic polyphenol, curcumin, in freely-moving rats. Food Chem. 127 (3), 918–925.

Tsai, Y.M., Chien, C.F., Lin, L.C., Tsai, T.H., 2011b. Curcumin and its nano-formulation: the kinetics of tissue distribution and blood-brain barrier penetration. Int. J. Pharm. 416, 331–338.

van der Merwe, C., van Dyk, H.C., Engelbrecht, L., van der Westhuizen, F.H., Kinnear, C., Loos, B., et al., 2017. Curcumin rescues a PINK1 knock down SH-SY5Y cellular model of Parkinson's disease from mitochondrial dysfunction and cell death. Mol. Neurobiol. 54, 2752–2762.

Wagner, S., Zensi, A., Wien, S.L., Tschickardt, S.E., Maier, W., Vogel, T., et al., 2012. Uptake mechanism of ApoE-modified nanoparticles on brain capillary endothelial cells as a blood-brain barrier model. PLoS One 7, e32568.

Wang, J., Du, X.X., Jiang, H., Xie, J.X., 2009. Curcumin attenuates 6-hydroxydopamine induced cytotoxicity by anti-oxidation and nuclear factor-kappa B modulation in MES23.5 cells. Biochem. Pharmacol. 78, 178–183.

Wang, X.S., Zhang, Z.R., Zhang, M.M., Sun, M.X., Wang, W.W., Xie, C.L., 2017a. Neuroprotective properties of curcumin in toxin-base animal models of Parkinson's disease: a systematic experiment literatures review. BMC Complement Altern. Med. 17, 412.

Wang, Y.L., Ju, B., Zhang, Y.Z., Yin, H.L., Liu, Y.J., Wang, S.S., et al., 2017b. Protective effect of curcumin against oxidative stress induced injury in rats with Parkinson's disease through the Wnt/β-catenin signaling pathway. Cell. Physiol. Biochem. 43, 2226–2241.

Wu, Q., Fang, J., Li, S., Wei, J., Yang, Z., Zhao, H., 2017. Interaction of bisphenol A 3,4-quinone metabolite with glutathione and ribonucleosides/deoxyribonucleosides *in vitro*. J. Hazard Mater. 323, 195–202.

Yang, D., Li, T., Liu, Z., Arbez, N., Yan, J., Moran, T.H., et al., 2012. LRRK2 kinase activity mediates toxic interactions between genetic mutation and oxidative stress in a Drosophila model: suppression by curcumin. Neurobiol. Dis. 47, 385–392.

Yang, J., Song, S., Li, J., Liang, T., 2014. Neuroprotective effect of curcumin on hippocampal injury in 6-OHDA-induced Parkinson's disease rat. Pathol. Res. Pract. 210, 357–362.

Yao, M.F., Xiao, H., McClements, D.J., 2014. Delivery of lipophilic bioactives: assembly, disassembly, and reassembly of lipid nanoparticles. Annu. Rev. Food Sci. Technol. 5, 53–81.

Young, N.A., Bruss, M.S., Gardner, M., Willis, W.L., Mo, X., Valiente, G.R., et al., 2014. Oral administration of nano-emulsion curcumin in mice suppresses inflammatory-induced NF kappa B signaling and macrophage migration. PLoS One 9, e111559.

Yu, S., Wang, Y., Wang, X., 2012. Curcumin prevents dopaminergic neuronal death in experimental Parkinson's disease research. J. China Med. Univ. 41, 569–570.

Yu, S., Wang, X., He, X., Wang, Y., Gao, S., Ren, L., et al., 2016. Curcumin exerts anti-inflammatory and antioxidative properties in 1-methyl-4-phenylpyridinium ion (MPP(+))-stimulated mesencephalic astrocytes by interference with TLR4 and downstream signaling pathway. Cell Stress Chaperones 21, 697–705.

Yue, X., Dai, Z., 2014. Recent advances in liposomal nanohybrid cerasomes as promising drug nanocarriers. Adv. Colloid Interface 207, 32–42.

Zbarsky, V., Datla, K.P., Parkar, S., Rai, D.K., Aruoma, O.I., Dexter, D.T., 2005. Neuroprotective properties of the natural phenolic antioxidants curcumin and naringenin but not quercetin and fisetin in a 6-ohda model of Parkinson's disease. Free Radic. Res. 39, 1119–1125.

Zhang, L.J., Wu, C.F., Meng, X.L., Yuan, D., Cai, X.D., Wang, Q.L., et al., 2008. Comparison of inhibitory potency of three different curcuminoid pigments on nitric oxide and tumor necrosis factor production of rat primary microglia induced by lipopolysaccharide. Neurosci. Lett. 447, 48–53.

Zhang, L., Wu, C., Zhao, S., Yuan, D., Lian, G., Wang, X., et al., 2010. Demethoxycurcumin, a natural derivative of curcumin attenuates LPS-induced pro-inflammatory responses through down-regulation of intracellular ROS-related MAPK/NF-kappaB signaling pathways in N9 microglia induced by lipopolysaccharide. Int. Immunopharmacol. 10, 331–338.

Zhang, N., Yan, F., Liang, X., Wu, M., Shen, Y., Chen, M., Xu, Y., Zou, G., Jiang, P., Tang, C., Zheng, H., Dai, Z., 2018. Localized delivery of curcumin into brain with polysorbate 80-modified cerasomes by ultrasound-targeted microbubble destruction for improved Parkinson's disease therapy. Theranostics 8, 2264–2277.

Zhao, C., Li, H., Zhao, X.J., Liu, Z.X., Zhou, P., Liu, Y., et al., 2016. Heat shock protein 60 affects behavioral improvement in a rat model of Parkinson's disease grafted with human umbilical cord mesenchymal stem cell-derived dopaminergic-like neurons. Neurochem. Res. 41, 1238–1249.

CHAPTER 6

Redox nanoparticles: the corner stones in the development of Parkinson's disease therapeutics

1. Introduction

Wealth of studies has shown the pivotal role of mitochondrial dysfunction and oxidative stress in neurodegenerative disorders such as Parkinson's disease (PD) (Mattson et al., 1999; Beal, 2003; Jenner, 2003; Gille et al., 2004; Schapira, 2007; Exner et al., 2012; Hauser and Hastings, 2012; Sikorska et al., 2014). A common pathological event found in several neurodegenerative diseases such as PD is neuroinflammation associated with generation of reactive oxygen species (ROS) and proinflammatory factors (McGeer et al., 1988; Perry et al., 1995; Busciglio and Yankner, 1995; Ebadi et al., 1996; Wu et al., 2003, Klyachko et al., 2014). To overcome ROS-induced oxidative stress, numerous antioxidants were extensively studied but they yielded limited success due to insufficient brain delivery of enzymes through blood—brain barrier (BBB) (Klyachko et al., 2014). With a view to overcoming this biological hurdle, a potential delivery system for redox enzyme catalase has been developed (Batrakova et al., 2007, Brynskikh et al., 2010, Zhao et al., 2011a,b, Klyachko et al., 2014). Accordingly, drug and antioxidant transport to the brain play a pivotal role in neurodegenerative disease therapeutics (Muzykantov, 2001, Supinski and Callahan, 2006, Thomas et al., 2008, Hood et al., 2011, Klyachko et al., 2014). Deterioration of antioxidant levels has been found in early stage PD such as deterioration of vital cellular antioxidant and glutathione in the *substantia nigra pars compacta* (SNpc) (Sian et al., 1994, Sadowska-Bartosz and Bartosz, 2018).

Increased iron levels were found in PD brains more specifically in the nigrostriatal dopaminergic system where redox pair generation between iron and dopamine was observed (Hare and Double, 2016, Sadowska-Bartosz and Bartosz, 2018). Perturbed copper levels were also found to contribute to PD

Parkinson's Disease Therapeutics
ISBN 978-0-12-819882-7
https://doi.org/10.1016/B978-0-12-819882-7.00006-4

(Davies et al., 2016, Sadowska-Bartosz and Bartosz, 2018). Transition metal–mediated toxicity has been implicated in the etiology of neurodegenerative diseases. Therefore, treatment with metal chelators was of considerable importance in both Alzheimer's disease (AD) and PD (Sadowska-Bartosz and Bartosz, 2018).

2. Parkinson's disease diagnosis

PD diagnosis has been a herculean task due to the lack of appropriate tests of biosensors (Ma et al., 2013). Overlapping symptoms in early stages of PD make the differential diagnosis more cumbersome (Jellinger, 2003, Litvan et al., 2007, Shi et al., 2011, Ma et al., 2013). Therefore, there is a growing need to develop potential biosensor to diagnose PD and monitor its progress accurately. Wealth of studies has shown that mitochondrial NADH:ubiquinone oxidoreductase (complex I) attenuation underlies the pathogenesis of sporadic PD, which induces dopamine neuronal death (Betarbet et al., 2000, Dawson and Dawson, 2003, Valente et al., 2004, Ma et al., 2013). Ubiquinone, which is also called coenzyme Q, is localized in the hydrophobic core of phospholipid bilayer of mitochondrial inner membrane. It acts as a mobile transporter of electrons and protons [Do et al., 1996, Matthews et al., 1998, Ma et al., 2013]. These redox molecules with charge transfer when conjugated with quantum dots (QDs) can be used as robust biosensor molecules (Chan and Nie, 1998, Melinger et al., 2010; Ma et al., 2013). Accordingly, in dopamine–QD conjugates, dopamine donated electrons sensitized QDs by multifarious mechanisms entailing ROS (Medintz et al., 2005; Clarke et al., 2006; Cooper et al., 2009, 2010; Ma et al., 2013). These characteristics of QDs have been explored to design biosensors to diagnose the levels of complex I, which in turn can diagnose PD.

3. Coenzyme Q10

Neurotoxic chemicals such as MPTP (1-methyl-4-phenyl-1,2,3,6-tetrahydropyridine), paraquat, and rotenone impede complex I of oxidative phosphorylation pathway and instigate disintegration of dopaminergic neurotransmission in rodents (Cannon et al., 2009; Cicchetti et al., 2005; Jackson-Lewis and Przedborski, 2007; McCarthy et al., 2004, Sikorska et al., 2014). In line with this, decreased levels of coenzyme Q10 (CoQ10) were observed in platelets and the brain of PD patients (Shults et al., 1998,

1999; Hargreaves et al., 2008; Sikorska et al., 2014). CoQ10 present primarily in membranes in the reduced form of quinol induces substantial antioxidant activity by reacting with ROS or revitalizing α-tocopherol and ascorbate (Crane, 2001; Turunen et al., 2004; Sikorska et al., 2014). Therefore, plethora of studies focused on antioxidants that significantly curtail ROS and associated oxidative stress (Koppula et al., 2012; Sikorska et al., 2014). Nevertheless, numerous antioxidants targeted against mitochondrial damage yielded limited success (Hart et al., 2009; Lew, 2011; Sikorska et al., 2014). Despite the remarkable therapeutic efficacy of CoQ10, inability to enter brain, low absorption in the body, and solubility limit its success (Sikorska et al., 2014). Therefore, there is a growing need to develop innovative formulation to overcome these challenges.

3.1 Nanotechnology

Nanoparticles (NPs) have long been used in nanomedicine because of their substantial rolein diagnosis and therapy of manifold diseases such as neurodegenerative diseases (Singh et al., 2008; Obulesu and Jhansilakshmi, 2016, Sadowska-Bartosz and Bartosz, 2018). Although NPs offer better therapeutic effects, yet their toxicity toward neurons, cellular counterparts, and BBB is worth considering factor (Cupaioli et al., 2014, Sadowska-Bartosz and Bartosz, 2018). Presence of sufficient surface area-to-volume ratio of NPs results in enhanced electron densities on the surface and offers adequate catalytic activity (Karakoti et al., 2010, Sadowska-Bartosz and Bartosz, 2018). In line with this, metal NPs render remarkable ROS scavenging and deoxygenating activities (Xu and Qu, 2014, Sadowska-Bartosz and Bartosz, 2018). Surprisingly, panoply of studies has designed nanozymes (NP-based artificial antioxidant enzymes), which substantially attenuate apoptosis and enhance cell survival (Wei and Wang, 2013, Sadowska-Bartosz and Bartosz, 2018).

3.2 Coenzyme Q10 delivery system

To overcome the abovementioned challenges and to improve solubilization of CoQ10, a self-emulsifying poly(ethylene glycol) (PEG)—obtained α-tocopherol (PTS) nanomicelles were designed. This biocompatible CoQ10/PTS formulation (Ubisol-Q10) with potential delivery ability combated oxidative stress or glutamate excitotoxicity instigated apoptosis in human NT2 and SH-SY5Y cells (Borowy-Borowski et al., 2004; Sikorska et al., 2003; McCarthy et al., 2004; Somayajulu et al., 2005; Sandhu et al.,

2003; Sikorska et al., 2014). Its oral administration also protected dopaminergic neurons of SNpc of rats treated with paraquat (Somayajulu-Nitu et al., 2009; Sikorska et al., 2014).

Nanomicellar preparation of CoQ10 (Ubisol-Q10) showed enhanced brain spanning and therapeutic efficacy against neurodegeneration in MPTP-treated mouse model (Sikorska et al., 2014). In addition, it also activated astrocytes emphasizing that astrocytes are important in inducing neuroprotection (Sikorska et al., 2014). While the oral administration of the formulation showed appreciable protection against neurodegeneration, its discontinuation leads to neurodegeneration (Sikorska et al., 2014).

3.3 Nanozyme

Nanoparticulate redox enzyme, catalase, in macrophages intensely inhibits oxidative stress and protects dopaminergic neurons in animal models of PD (Klyachko et al., 2014). Cross-linked nanozyme encapsulated in macrophages profoundly decreased neuroinflammatory responses and enhanced neuronal protection in mice (Klyachko et al., 2014). To guard the catalase enzyme in macrophages, in this study, it has been encapsulated in block ionomer complexes containing cationic block copolymer, poly(ethyl eneimine) (PEI) and PEG (PEI-PEG), known as nanozyme (Zhao et al., 2011a,b, Klyachko et al., 2014).

3.4 Dopamine nanoparticles

To compensate the shortage of dopamine in PD, dopamine supplementation is done in the form of L-DOPA (L-3,4-dihydroxyphenylalanine), dopamine receptor agonists, inhibitors of dopamine disintegrating catechol-o-methyltransferase, and monoamine oxidase B either unaided or in association because the dopamine by itself cannot ferry BBB (Salat and Tolosa, 2013; Connolly and Lang, 2014; Krishna et al., 2014; Pahuja et al., 2015). Polylactic-co-glycolic acid (PLGA) has been extensively used to deliver pharmacological compounds to the brain because of its biocompatibility (D'Aurizio et al., 2011; Ren et al., 2011; Wang et al., 2012; Regnier-Delplace et al., 2013; Pahuja et al., 2015). It is biodegradable, and after degradation, lactic acid and glycolic acid are removed as carbon dioxide and water through the Krebs cycle.

Dopamine-loaded PLGA NPs showed protective efficacy in 6-OHDA-treated parkinsonian rats by attenuating ROS production, dopaminergic neuron degeneration, and dopamine autoxidation-mediated toxicity

(Pahuja et al., 2015). These NPs showed controlled release of dopamine in the brain. Because continuous and controlled delivery of dopamine is very essential in PD, these PLGA NPs are of utmost importance in therapy (Carlsson et al., 2005; Pellicano et al., 2013; Wright and Waters, 2013; Pahuja et al., 2015). Interestingly, PLGA NPs profoundly protected dopamine against oxidation (Pahuja et al., 2015). Mounting evidence has shown the internalization of PEI-PLGA NPs in neuronal cells (Park et al., 2007; Singhal et al., 2013; Pahuja et al., 2015). The underlying mechanism of PLGA NPs internalization was found to be clathrin-mediated endocytosis (Hu et al., 2011; Pahuja et al., 2015).

3.4.1 Redox-active nanoparticles

Multifarious redox-active nanoparticles (RNPs) such as cerium oxide RNPs, boron cluster containing and silica containing Gd3 N@C80 showed significant amelioration in both AD and PD cellular and animal models (Sadowska-Bartosz and Bartosz, 2018).

3.4.2 Cerium oxide nanoparticles

Cerium, a highly reactive lanthanide, occurs in Ce^{3+} and Ce^{4+}, but Ce^{4+} is more stable (Xu and Qu, 2014, Sadowska-Bartosz and Bartosz, 2018). Accordingly, it occurs in two oxides namely cerium dioxide (CeO_2) and cerium sesquioxide (Ce_2O_3), and CeO_2 is more stable (Xu and Qu, 2014, Sadowska-Bartosz and Bartosz, 2018). Methods used in the synthesis of cerium oxide NPs are thermal decomposition, solvothermal oxidation, microwave-associated solvothermal process, sol—gel microemulsion, and flame spray pyrolysis (Xu et al., 2008, Sadowska-Bartosz and Bartosz, 2018).

CeO_2 NPs exhibiting catalase, superoxide dismutase activity, or peroxidase activity have been studied (Pirmohamed et al., 2010, Sadowska-Bartosz and Bartosz, 2018). CeO_2 NPs conjugated to PEG and of 3 nm size were found to ferry through BBB and curtail oxidative stress in Sprague—Dawley rats (Gao et al., 2014; Bailey et al., 2016; Sack-Zschauer et al., 2017, Sadowska-Bartosz and Bartosz, 2018).

4. Conclusions and future perspectives

Redox NPs exhibit profound antioxidant effect by mimicking the potential antioxidant enzymes. They play an essential role in scavenging ROS and defending tissues from toxicity. Redox NPs showed remarkably high protective efficacy against PD compared to other drug delivery systems. Although Ubisol-Q10 showed significant protection against MPTP-induced toxicity in rats, when discontinued, neurodegeneration was initiated. In addition, there is little understanding of this compound's brain spanning (Sikorska et al., 2014). CeO_2 NPs play a significant role in ameliorating PD symptoms by spanning the BBB effectively. Plethora of NPs developed using a wide range of ligands to pass through BBB yielded worthy results. However, several systemic toxicity issues remain to be solved. Therefore, more appropriate delivery systems have been warranted.

References

Bailey, Z.S., Nilson, E., Bates, J.A., Oyalowo, A., Hockey, K.S., Sajja, V.S.S.S., et al., 2016. Cerium oxide nanoparticles improve outcome after in vitro and in vivo mild traumatic brain injury. J. Neurotrauma 33, 1–11.

Batrakova, E.V., Li, S., Reynolds, A.D., Mosley, R.L., Bronich, T.K., Kabanov, A.V., et al., 2007. A macrophage-nanozyme delivery system for Parkinson's disease. Bioconjug. Chem. 18, 1498–1506.

Beal, M.F., 2003. Mitochondria, oxidative damage, and inflammation in Parkinson's disease. Ann. N. Y. Acad. Sci. 991, 120–131.

Betarbet, R., Sherer, T.B., MacKenzie, G., Garcia-Osuna, M., Panov, A.V., Greenamyre, J.T., 2000. Chronic systemic pesticide exposure reproduces features of Parkinson's disease. Nat. Neurosci. 3, 1301–1306.

Borowy-Borowski, H., Sodja, C., Docherty, J., Walker, P.R., Sikorska, M., 2004. Unique technology for solubilization and delivery of highly lipophilic bioactive molecules. J. Drug Target. 12, 415–424.

Brynskikh, A.M., Zhao, Y., Mosley, R.L., Li, S., Boska, M.D., Klyachko, N.L., et al., 2010. Macrophage delivery of therapeutic nanozymes in a murine model of Parkinson's disease. Nanomedicine (Lond.) 5, 379–396.

Busciglio, J., Yankner, B.A., 1995. Apoptosis and increased generation of reactive oxygen species in Down's syndrome neurons *in vitro*. Nature 378, 776–779.

Cannon, J.R., Tapias, V., Na, H.M., Honick, A.S., Drolet, R.E., Greenamyre, J.T., 2009. A highly reproducible rotenone model of Parkinson's disease. Neurobiol. Dis. 34, 279–290.

Carlsson, T., Winkler, C., Burger, C., Muzyczka, N., Mandel, R.J., Cenci, A., et al., 2005. Reversal of dyskinesias in an animal model of Parkinson's disease by continuous L-DOPA delivery using rAAV vectors. Brain 128, 559–569.

Chan, W.C.W., Nie, S., 1998. Quantum dot bioconjugates for ultrasensitive nonisotopic detection. Science 281, 2016–2018.

Cicchetti, F., Lapointe, N., Roberge-Tremblay, A., Saint-Pierre, M., Jimenez, L., Ficke, B.W., et al., 2005. Systemic exposure to paraquat and maneb models early Parkinson's disease in young adult rats. Neurobiol. Dis. 20, 360–371.

Clarke, S.J., Hollmann, C.A., Zhang, Z., Suffern, D., Bradforth, S.E., Dimitrijevic, N.M., et al., 2006. Photophysics of dopamine-modified quantum dots and effects on biological systems. Nat. Mater. 5, 409–417.

Connolly, B.S., Lang, A.E., 2014. Pharmacological treatment of Parkinson disease: a review. JAMA J. Am. Med. Assoc. 311, 1670–1683.

Cooper, D.R., Suffern, D., Carlini, L., Clarke, S.J., Parbhoo, R., Bradforth, S.E., et al., 2009. Photoenhancement of lifetimes in CdSe/ZnS and CdTe quantum dot-dopamine conjugates. Phys. Chem. Chem. Phys. 11, 4298–4310.

Cooper, D.R., Dimitrijevic, N.M., Nadeau, J.L., 2010. Photosensitization of CdSe/ZnS QDs and reliability of assays for reactive oxygen species production. Nanoscale 2, 114–121.

Crane, F.L., 2001. Biochemical functions of coenzyme Q10. J. Am. Coll. Nutr. 20, 591–598.

Cupaioli, F.A., Zucca, F.A., Boraschi, D., Zecca, L., 2014. Engineered nanoparticles. How brain friendly is this new guest? Prog. Neurobiol. 119–120, 20–38.

D'Aurizio, E., Sozio, P., Cerasa, L.S., Vacca, M., Brunetti, L., Orlando, G., et al., 2011. Biodegradable microspheres loaded with an anti-Parkinson prodrug: an in vivo pharmacokinetic study. Mol. Pharm. 8, 2408–2415.

Davies, K.M., Mercer, J.F., Chen, N., Double, K.L., 2016. Copper dyshomeostasis in Parkinson's disease: implications for pathogenesis and indications for novel therapeutics. Clin. Sci. 130, 565–574.

Dawson, T.M., Dawson, V.L., 2003. Molecular pathways of neurodegeneration in Parkinson's disease. Science 302, 819–822.

Do, T.Q., Schultz, J.R., Clarke, C.F., 1996. Enhanced sensitivity of biquinone deficient mutants of *Saccharomyces cerevisiae* to products of autoxidized polyunsaturated fatty acids. Proc. Natl. Acad. Sci. U.S.A. 93, 7534–7539.

Ebadi, M., Srinivasan, S.K., Baxi, M.D., 1996. Oxidative stress and antioxidant therapy in Parkinson's disease. Prog. Neurobiol. 48, 1–19.

Exner, N., Lutz, A.K., Haass, C., Winklhofer, K.F., 2012. Mitochondrial dysfunction in Parkinson's disease: molecular mechanisms and pathophysiological consequences. EMBO J. 31, 3038–3062.

Gao, Y., Chen, K., Ma, J.L., Gao, F., 2014. Cerium oxide nanoparticles in cancer. OncoTargets Ther. 7, 835–840.

Gille, G., Hung, S.T., Reichmann, H., Rausch, W.D., 2004. Oxidative stress to dopaminergic neurons as models of Parkinson's disease. Ann. N. Y. Acad. Sci. 1018, 533–540.

Hare, D.J., Double, K.L., 2016. Iron and dopamine: a toxic couple. Brain 139, 1026–1035.

Hargreaves, I.P., Lane, A., Sleiman, P.M., 2008. The coenzyme Q10 status of the brain regions of Parkinson's disease patients. Neurosci. Lett. 447, 17–19.

Hart, R.G., Pearce, L.A., Ravina, B.M., Yaltho, T.C., Marler, J.R., 2009. Neuroprotection trials in Parkinson's disease: systematic review. Mov. Disord. 24, 647–654.

Hauser, D.N., Hastings, T.G., 2012. Mitochondrial dysfunction and oxidative stress in Parkinson's disease and monogenic parkinsonism. Neurobiol. Dis. 51, 35–42.

Hood, E., Simone, E., Wattamwar, P., Dziubla, T., Muzykantov, V., 2011. Nanocarriers for vascular delivery of antioxidants. Nanomedicine (Lond.) 6, 1257–1272.

Hu, K., Shi, Y., Jiang, W., Han, J., Huang, S., Jiang, X., 2011. Lactoferrin conjugated PEG-PLGA nanoparticles for brain delivery: preparation, characterization and efficacy in Parkinson's disease. Int. J. Pharm. 415, 273–283.

Jackson-Lewis, V., Przedborski, S., 2007. Protocol for the MPTP mouse model of Parkinson's disease. Nat. Protoc. 2, 141–151.

Jellinger, K.A., 2003. Neuropathological spectrum of synucleinopathies. Mov. Disord. 18, S2–S12.

Jenner, P., 2003. Oxidative stress in Parkinson's disease. Ann. Neurol. 53, S26–S36.

Karakoti, A., Singh, S., Dowding, J.M., Seal, S., Self, W.T., 2010. Redox-active radical scavenging nanomaterials. Chem. Soc. Rev. 39, 4422–4432.

Klyachko, N.L., Haney, M.J., Zhao, Y., Manickam, D.S., Mahajan, V., Suresh, P., et al., 2014. Macrophages offer a paradigm switch for CNS delivery of therapeutic proteins. Nanomedicine (Lond) 9, 1403–1422.

Koppula, S., Kumar, H., More, S.V., Kim, B.W., Kim, I.S., Choi, D.K., 2012. Recent advances on the neuroprotective potential of antioxidants in experimental models of Parkinson's disease. Int. J. Mol. Sci. 13, 10608–10629.

Krishna, R., Ali, M., Moustafa, A.A., 2014. Effects of combined MAO-B inhibitors and Levodopa vs. Monotherapy in Parkinson's disease. Front. Aging Neurosci. 6, 180.

Lew, M.F., 2011. The evidence for disease modification in Parkinson's disease. Int. J. Neurosci. 121, 18–26.

Litvan, I., Halliday, G., Hallett, M., Goetz, C.G., Rocca, W., Duyckaerts, C., et al., 2007. The etiopathogenesis of Parkinson disease and suggestions for future research. Part I. J. Neuropathol. Exp. Neurol. 66, 251–257.

Ma, W., Qin, L.X., Liu, F.T., Gu, Z., Wang, J., Pan, Z.G., et al., 2013. Ubiquinone-quantum dot bioconjugates for in vitro and intracellular complex I sensing. Sci. Rep. 3, 1537.

Matthews, R.T., Yang, L., Browne, S., Baik, M., Beal, M.F., 1998. Coenzyme Q10 administration increases brain mitochondrial concentrations and exerts neuroprotective effects. Proc. Natl. Acad. Sci. U.S.A. 95, 8892–8897.

Mattson, M.P., Pedersen, W.A., Duan, W., Culmsee, C., Camandola, S., 1999. Cellular and molecular mechanisms underlying perturbed energy metabolism and neuronal degeneration in Alzheimer's and Parkinson's diseases. Ann. N. Y. Acad. Sci. 893, 154–175.

McCarthy, S., Somayajulu, M., Sikorska, M., Borowy-Borowski, H., Pandey, S., 2004. Paraquat induces oxidative stress and neuronal cell death; neuroprotection by water-soluble coenzyme Q10. Toxicol. Appl. Pharmacol. 201, 21–31.

McGeer, P.L., Itagaki, S., Boyes, B.E., McGeer, E.G., 1988. Reactive microglia are positive for HLA-DR in the substantia nigra of Parkinson's and Alzheimer's disease brains. Neurology 38, 1285–1291.

Medintz, I.L., Tetsuouyeda, H., Goldman, E.R., Mattoussi, H., 2005. Quantum dot bioconjugates for imaging, labelling and sensing. Nat. Mater. 4, 435–446.

Melinger, J.S., Blanco-Canosa, J.B., Dawson, P.E., Mattoussi, H., 2010. Quantum-dot/dopamine bioconjugates function as redox coupled assemblies for in vitro and intracellular pH sensing. Nat. Mater. 9, 676–684.

Muzykantov, V.R., 2001. Targeting of superoxide dismutase and catalase to vascular endothelium. J. Control. Release 71, 1–21.

Obulesu, M., Jhansilakshmi, M., 2016. Neuroprotective role of nanoparticles against Alzheimer's disease. Curr. Drug Metabol. 17, 142–149.

Pahuja, R., Seth, K., Shukla, A., Shukla, R.K., Bhatnagar, P., Chauhan, L.K., et al., 2015. Trans-blood brain barrier delivery of dopamine-loaded nanoparticles reverses functional deficitsin parkinsonian rats. ACS Nano 9, 4850–4871.

Park, I.K., Lasiene, J., Chou, S.H., Horner, P.J., Pun, S.H., 2007. Neuron-specific delivery of nucleic acids mediated by tet1-modified poly(ethylenimine). J. Gene Med. 9, 691–702.

Pellicano, C., Benincasa, D., Fanciulli, A., Latino, P., Giovannelli, M., Pontieri, F.E., 2013. The impact of extended release dopamine agonists on prescribing patterns for therapy of early Parkinson's disease: an observational study. Eur. J. Med. Res. 18–60.

Perry, V.H., Bell, M.D., Brown, H.C., Matyszak, M.K., 1995. Inflammation in the nervous system. Curr. Opin. Neurobiol. 5, 636–641.

Pirmohamed, T., Dowding, J.M., Singh, S., Wasserman, B., Heckert, E., Karakoti, A.S., et al., 2010. Nanoceria exhibit redox state-dependent catalase mimetic activity. Chem. Commun. 46, 2736–2738.

Regnier-Delplace, C., Thillaye du Boullay, O., Siepmann, F., Martin-Vaca, B., Degrave, N., Demonchaux, P., et al., 2013. PLGA microparticles with zero-order release of the Labile anti-Parkinson drug apomorphine. Int. J. Pharm. 443, 68–79.

Ren, T., Yang, X., Wu, N., Cai, Y., Liu, Z., Yuan, W., 2011. Sustained-release formulation of Levodopa methyl ester/benserazide for prolonged suppressing dyskinesia expression in 6-OHDA-leisoned rats. Neurosci. Lett. 502, 117–122.

Sack-Zschauer, M., Karaman-Aplak, E., Wyrich, C., Das, S., Schubert, T., Meyer, H., et al., 2017. Efficacy of different compositions of cerium oxide nanoparticles in tumor–stroma interaction. J. Biomed. Nanotechnol. 13, 1735–1746.

Sadowska-Bartosz, I., Bartosz, G., 2018. Redox nanoparticles: synthesis, properties and perspectives of use for treatment of neurodegenerative diseases. J. Nanobiotechnol. 16, 87.

Salat, D., Tolosa, E., 2013. Levodopa in the treatment of Parkinson's disease: current status and new developments. J. Parkinson's Dis. 3, 255–269.

Sandhu, J.K., Pandey, S., Ribecco, M., Monette, R., Borowy-Borowski, H., Walker, P.R., et al., 2003. Molecular mechanisms of glutamate neurotoxicity in mixed cultures of NT2-derived neurons and astrocytes: protective effects of coenzyme Q10. J. Neurosci. Res. 72, 691–703.

Schapira, A.H., 2007. Mitochondrial dysfunction in Parkinson's disease. Cell Death Differ. 14, 1261–1266.

Shi, M., Bradner, J., Hancock, A.M., Chung, K.A., Quinn, J.F., Peskind, E.R., et al., 2011. Cerebrospinal fluid biomarkers for Parkinson disease diagnosis and progression. Ann. Neurol. 69, 570–580.

Shults, C.W., Beal, M.F., Fontaine, D., Nakano, K., Haas, R.H., 1998. Absorption, tolerability, and effects on mitochondrial activity of oral coenzyme Q10 in parkinsonian patients. Neurology 50, 793–795.

Shults, C.W., Haas, R.H., Beal, M.F., 1999. A possible role of coenzyme Q10 in the etiology and treatment of Parkinson's disease. Biofactors 9, 267–272.

Sian, J., Dexter, D.T., Lees, A.J., Daniel, S., Agid, Y., Javoy-Agid, F., et al., 1994. Alterations in glutathione levels in Parkinson's disease and other neurodegenerative disorders affecting basal ganglia. Ann. Neurol. 36, 348–355.

Sikorska, M., Borowy-Borowski, H., Zurakowski, B., Walker, P.R., 2003. Derivatised alpha-tocopherol as a CoQ10 carrier in a novel water-soluble formulation. Biofactors 18, 173–183.

Sikorska, M., Lanthier, P., Miller, H., Beyers, M., Sodja, C., Zurakowski, B., et al., 2014. Nanomicellar formulation of coenzyme Q10 (Ubisol-Q10) effectively blocks ongoingneurodegeneration in the mouse 1-methyl-4-phenyl-1,2,3,6-tetrahydropyridine model: potential use as an adjuvant treatment in Parkinson's disease. Neurobiol. Aging 35, 2329–2346.

Singh, M., Singh, S., Prasad, S., Gambhir, I.S., 2008. Nanotechnology in medicine and antibacterial effect of silver nanoparticles. Digest. J. Nanomater. Biostruct. 3, 115–122.

Singhal, A., Morris, V.B., Labhasetwar, V., Ghorpade, A., 2013. Nanoparticle-mediated catalase delivery protects human neurons from oxidative stress. Cell Death Dis. 4, e903.

Somayajulu, M., McCarthy, S., Hung, M., Sikorska, M., Borowy-Borowski, H., Pandey, S., 2005. Role of mitochondria in neuronal cell death induced by oxidative stress; neuroprotection by coenzyme Q10. Neurobiol. Dis. 18, 618–627.

Somayajulu-Nitu, M., Sandhu, J.K., Cohen, J., Sikorska, M., Sridhar, T.S., Matei, A., et al., 2009. Paraquat induces oxidative stress, neuronal loss in substantia nigra region and parkinsonism in adult rats: neuroprotection and amelioration of symptoms by water-soluble formulation of coenzyme Q10. BMC Neurosci. 10, 88.

Supinski, G.S., Callahan, L.A., 2006. Polyethylene glycol-superoxide dismutase prevents endotoxin-induced cardiac dysfunction. Am. J. Respir. Crit. Care Med. 173, 1240–1247.

Thomas, S.R., Witting, P.K., Drummond, G.R., 2008. Redox control of endothelial function and dysfunction: molecular mechanisms and therapeutic opportunities. Antioxidants Redox Signal. 10, 1713–1765.

Turunen, M., Olsson, J., Dallner, G., 2004. Metabolism and function of coenzyme Q. Biochim. Biophys. Acta 1660, 171–199.

Valente, E.M., Abou-Sleiman, P.M., Caputo, V., Muqit, M.M., Harvey, K., Gispert, S., et al., 2004. Hereditary early-onset Parkinson's disease caused by mutations in PINK1. Science 304, 1158–1160.

Wang, A., Wang, L., Sun, K., Liu, W., Sha, C., Li, Y., 2012. Preparation of rotigotine-loaded microspheres and their combination use with L-DOPA to modify dyskinesias in 6-OHDA lesioned rats. Pharm. Res. 29, 2367–2376.

Wei, H., Wang, E., 2013. Nanomaterials with enzyme-like characteristics (nanozymes): next-generation artificial enzymes. Chem. Soc. Rev. 42, 6060–6093.

Wright, B.A., Waters, C.H., 2013. Continuous dopaminergic delivery to minimize motor complications in Parkinson's disease. Expert Rev. Neurother. 13, 719–729.

Wu, D.C., Teismann, P., Tieu, K., Vila, M., Jackson-Lewis, V., Ischiropoulos, H., et al., 2003. NADPH oxidase mediates oxidative stress in the 1-methyl-4-phenyl-1,2,3,6-tetrahydropyridine model of Parkinson's disease. Proc. Natl. Acad. Sci. U.S.A. 100, 6145–6150.

Xu, C., Qu, X., 2014. Cerium oxide nanoparticle: a remarkably versatile rare earth nanomaterial for biological applications. NPG Asia Mater. 6, e90.

Xu, J., Li, G., Li, L., 2008. CeO_2 nanocrystals: seed-mediated synthesis and size control. Mater. Res. Bull. 43, 990–995.

Zhao, Y., Haney, M.J., Klyachko, N.L., Li, S., Booth, S.L., Higginbotham, S.M., et al., 2011a. Polyelectrolyte complex optimization for macrophage delivery of redox enzyme nanoparticles. Nanomedicine (Lond.) 6, 25–42.

Zhao, Y., Haney, M.J., Mahajan, V., Reiner, B.C., Dunaevsky, A., Mosley, R.L., et al., 2011b. Active targeted macrophage-mediated delivery of catalase to affected brain regions in models of Parkinson's disease. J. Nanomed. Nanotechnol. S4, 003.

CHAPTER 7

Liposomal maneuvers against Parkinson's disease

1. Introduction

1.1 Parkinson's disease

Parkinson's disease (PD) is the second most common neurodegenerative disease (ND) initially found in substantia nigra and characterized by bradykinesia and tremors (Poewe et al., 2017, Kang et al., 2018). Diminished dopamine level due to low dopamine transporters also results in considerable loss of neuronal functions (Nutt et al., 2004, Kang et al., 2018). Additionally, α-synuclein aggregation in the Lewy bodies is a potential marker of PD, but the molecular underpinnings are elusive (Spillantini et al., 1997, Kang et al., 2018). Although levodopa (L-DOPA) is extensively used drug for PD, there is no complete cure till date (Md et al., 2011; Agid et al., 1997, Kang et al., 2018). In addition, nonspecific targeting of L-DOPA induces cardiovascular side effects and dyskinesia (Thanvi et al., 2007, Kang et al., 2018).

2. Blood—Brain Barrier

Blood—brain barrier (BBB) has long been considered a primary impediment in the drug targeting to the brain and accomplishing significant therapeutic effect. Although a few pathways such as paracellular, transcellular, and carrier-mediated transport (Chapter 3, Fig. 7.1) have been employed by the neurotherapeutics, yet 98% small molecules and almost all large molecules fail to ferry BBB (Cetin et al., 2017; Teleanu et al., 2018). P-glycoprotein (P-gp) present in endothelial cells primarily attenuates drug transport through BBB (Wang et al., 2018). It being an adenosine triphosphate—dependent efflux pump effluxes diverse lipophilic compounds and (>400 Da) hydrophilic compounds (Ballabh et al., 2004, Wang et al., 2018). Therefore, a few small lipophilic molecules such as oxygen, carbon dioxide, and nicotine only can span BBB (Wang et al., 2018). Despite the

Parkinson's Disease Therapeutics
ISBN 978-0-12-819882-7
https://doi.org/10.1016/B978-0-12-819882-7.00007-6

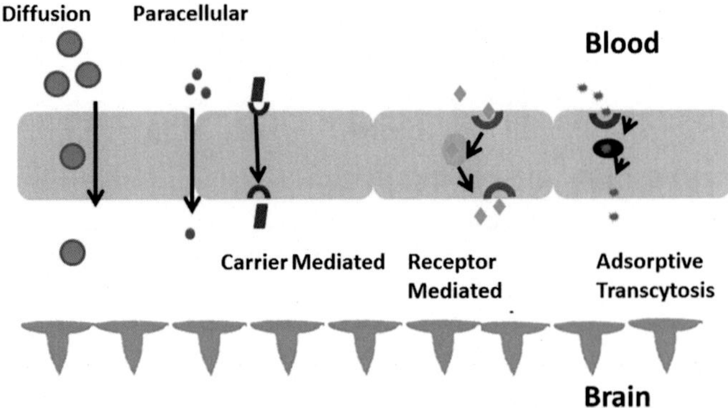

Figure 7.1 Multifarious pathways of nanoparticle entry through bllood brain barrier.

degenerated BBB in NDs, drug translocation across it is difficult because of the presence of P-gp and downregulation of pinocytosis (Zlokovic 2008; Wang et al., 2018). Therefore, manifold noninvasive nanotechnological methods have been used to overcome BBB and improve therapeutic delivery currently (Fonseca-Santos et al., 2015, Abou el Eloy et al., 2017, Davoodi and Saghavaz, 2017, Pignatello et al., 2017, Dong, 2018, Teleanu et al., 2018). Wealth of studies has concentrated on design of drug delivery systems (DDS) such as nanoparticles, dendrimers, liposomes, carbon nanotubes, and micelles for targeting therapeutic compounds, proteins, peptides, nucleic acids, or vaccines to the brain (Teleanu et al., 2018).

3. Liposomes

Liposomes are artificial spherical cells with single amphiphilic lipid bilayers and ability to encapsulate therapeutic molecules such as drugs, nucleic acids, proteins, and vaccines (Abbina and Parambath, 2018; Teleanu et al., 2018). Primarily they have hydrophilic core and one or more hydrophobic spaces bordered by lipid bilayers and can encapsulate both hydrophobic and hydrophilic compounds (Kang et al., 2018). Since their discovery in 1960, they have been widely used delivery systems for brain targeting (Vieira and Gamarra, 2016; Wang et al., 2018). They are self-assembling nanosystems that can encapsulate and release multifarious drugs to the central nervous system such as lipophilic drugs, hydrophilic drugs, protein centered drugs, and nucleic acid components (Agrawal et al., 2017; Khan et al., 2018).

Although they are considered as foreign elements by the biological system, they are biodegradable, nonimmunogenic, nonthrombogenic, and noncarcinogenic in nature (Gregoriadis, 2008, Siddiq et al., 2018). They show substantial efficacy of binding to biological membranes, loading multiple components and evading enzymatic degradation (Spuch and Navarro, 2011, Siddiqet al., 2018).

Liposomal formulations render appreciable advantages such as morphological characterization and various size distributions, surface functionalization by targeting moieties and surfactants, plethora of administration pathways, and stimuli-mediated drug release (Wakaskar, 2017; Khan et al., 2018). They were also found to overcome issues such as low solubility, low bioavailability, and accomplish targeted delivery to the brain (Lai et al., 2013; Loureiro et al., 2015; Wang et al., 2018). Liposomes play a pivotal role in the treatment of plethora of diseases such as cancer and NDs (Obulesu and Jhansilakshmi, 2016, Obulesu, 2019). Among the multifarious DDS in use currently, lipid-based nanocarriers such as liposomes and solid lipid nanoparticles (SLNs) and micelles have been extensively studied (Khan et al., 2018). Stimuli-sensitive liposomes, which respond to diverse changes in magnetic field, temperature, pH, and ultrasound intensity, have also been in use (Nair and Saiyed, 2011, Vieira and Gamarra, 2016). In addition, multifunctional theranostic liposomes have been designed, which can diagnose the disease and also help the therapy (Ramos-Cabrer and Campos, 2013; Vieira and Gamarra, 2016). In a recent study, liposomes loaded with quantum dot and apomorphine have spanned BBB and exerted therapeutic effect against brain cancer (Kelkar and Reineke, 2011; Wen et al., 2012; Vieira and Gamarra, 2016). Nanoliposomes being spherical forms of lipid bilayers are potential nanocarriers that can encapsulate drugs, genes, and imaging agents for theranostic applications in NDs (Jain et al., 1998; Xiang et al., 2012a,b, Kang et al., 2018).

4. Surface functionalization in brain targeting

The key factor responsible to enhance BBB entry of liposomes is the surface charge (neutral, positive, and negative) (Aharon et al., 2011; Wang et al., 2018). Positively charged liposomes were found to show low brain targeting efficacy compared to negative and neutral liposomes due to their capacity to bind proteins (Lockman et al., 2004; Campos-Martorell et al., 2016; Wang et al., 2018).

Liposomes are conjugated to diverse molecules such as poly(ethylene glycol) (PEG), transferrin, and glucose-vitamin C complex to accomplish enhanced systemic circulation of the formulations, successful BBB spanning, and enhanced aggregation at the target site (Table 7.1, Hu et al., 2017; Peng et al., 2018; Teleanu et al., 2018; Lakkadwala and Singh, 2018, 2019). To overcome BBB multifarious targeting moieties such as surfactants, transferrin, insulin, and cell-penetrating peptide (CPP) were used (Table 7.1, Lai et al., 2013; Kristensen and Brodin, 2017; Khan et al., 2018). Transferrin-coated liposomes were used to deliver α-mangostin, a substantial therapeutic agent in the treatment of NDs (Chen et al., 2016a,b; Teleanu et al., 2018). Liposomes tethered to monoclonal antibodies of glial fibrillary acidic protein showed enhanced BBB transport. In another study, a liposomal nanohybrid cerasome functionalized with polysorbate 80, a P-gp inhibitor, showed enhanced BBB transport of curcumin and considerable therapeutic effect against PD (Nisi Zhang et al., 2018, Wang et al., 2018).

4.1 Liposomes in Parkinson's disease

Liposomes encapsulated with N-3,4-bis(pivaloyloxy)-dopamine and functionalized with rabies virus glycoprotein 29 (RVG29) were designed to enhance the efficacy of dopamine (Khan et al., 2018). CPP-tethered liposomes were found to span BBB successfully, spread in specific brain regions, and enhance therapeutic efficacy in a PD mouse model (Qu et al., 2018a,b; Khan et al., 2018). Liposomal formulation profoundly enhanced the therapeutic effect of pharmaceutical compounds in PD mouse model (Qu et al., 2018a,b; Khan et al., 2018).

4.2 Liposomes in dopamine delivery

They are essentially used to deliver small hydrophobic molecules (SHM) such as dopamine and its derivatives to treat PD (Cao et al., 2016; Wang et al., 2018; Lopalco et al., 2018; Qu et al., 2018a,b). Growing lines of evidence have shown the remarkable therapeutic efficacy of SHM such as dopamine, L-DOPA, and rivastigmine in both in vitro and in vivo models of PD (Moretti et al., 2007, Ay et al., 2017; Wang et al., 2018). Chitosan, a cationic polyelectrolyte usually used to fabricate microcapsules, films, and

Table 7.1 List of various brain targeted liposomes.

Liposome	Conjugated ligand	Loaded therapeutic	Therapeutic effect	Reference
Liposome	Poly(ethylene glycol) Transferrin Glucose-vitamin C complex Insulin Cell-penetrating peptide		Enhanced systemic circulation and BBB spanning	Hu et al. (2017), Peng et al. (2018), Teleanu et al. (2018), Lakkadwala and Singh (2018), 2019 Lai et al. (2013), Kristensen and Brodin (2017), Khan et al. (2018)
	Transferrin	α-mangostin	Enhanced therapeutic efficacy	Chen et al. (2016a,b), Teleanu et al. (2018)
Liposomal nanohybrid cerasome	Polysorbate 80	Curcumin	Enhanced BBB transport of curcumin	Nisi Zhang, 2018, Wang et al. (2018)
Rabies virus glycoprotein 29 (RVG29) liposomes	RVG29	BPD	Enhanced efficacy of dopamine	Khan et al. (2018)
Liposomal L-DOPA	Chitosan	L-DOPA	Enhanced neuroprotection in PD	Cao et al. (2016), Wang et al. (2018)
DA-HCL-LPs	Transferrin	DA-HCL	Enhanced BBB spanning	Wang et al. (2018)
Maltodextrin liposomes	Maltodextrin	L-DOPA and glutathione co-loading	Enhanced BBB spanning	Gurturk et al. (2017)

Continued

Table 7.1 List of various brain targeted liposomes.—cont'd

Liposome	Conjugated ligand	Loaded therapeutic	Therapeutic effect	Reference
Chlorotoxin liposomes	Chlorotoxin	L-DOPA	Enhanced levodopa concentration in substantia nigra and striatum	Xiang et al. (2012a,b), Vieira and Gamarra (2016)
N-trimethyl chitosan chloride (TMC) liposomes	TMC	Curcumin	Decreased membrane dripping and enhanced loading efficiency	Chen et al. (2012), Wang et al. (2018)
Proteoliposomes	SUMO	rhFGF20	Enhanced BBB transport and dopaminergic neuron protection	Niu et al. (2018)

gels when associated with liposomal L-DOPA, showed neuroprotective efficacy in PD (Cao et al., 2016; Wang et al., 2018). Numerous cationic liposomes have been used to deliver therapeutic drugs and genes (Dass and Choong, 2006, Karmali and Chaudhuri, 2007, Chen et al., 2016a,b, Vieira and Gamarra, 2016). Lopalco et al. (2018) reported that transferrin-tethered liposomes loaded with dopamine hydrochloride (DA-HCL-LPs) showed profoundly improved BBB translocation in in vitro BBB model (Wang et al., 2018). In another study, maltodextrin-conjugated liposomes co-loaded with L-DOPA and glutathione (an antioxidant) showed significantly enhanced BBB permeation in Madin-Darby Canine Kidney(MDCK) cells (Gurturk et al., 2017). They are transported through receptor-mediated endocytosis. According to earlier studies, antioxidants avert phospholipid oxidation and enhance bioavailability of liposomes (Senior, 1987; Sies 1999; Gurturk et al., 2017). Accordingly, glutathione is co-loaded in this study to accomplish enhanced stability of liposomes (Sian et al., 1994; Shrestha et al., 2015; Gurturk et al., 2017). Chlorotoxin-conjugated liposomes loaded with L-DOPA enhanced levodopa concentration in substantia nigra and striatum and rendered significantly better therapeutic efficacy (Xiang et al., 2012a,b; Vieira and Gamarra, 2016). Accordingly, numerous anti-PD drug loaded liposomes ameliorated the dopamine release in striatum region (Yurasov et al., 1996, 1997; Di Stefano et al., 2004, 2006; Xiang et al., 2012a,b; Vieira and Gamarra, 2016).

4.3 Liposomes for natural compounds

4.3.1 Curcumin

Despite the profound antioxidant, antiinflammatory, and neuroprotective efficacy of curcumin against NDs, low solubility, bioavailability, and short systemic circulation limit its success (Chapter 5, Cole et al., 2007; Darvesh et al., 2012; Wang et al., 2018). It showed a few beneficial roles such as attenuation of α-synuclein accumulation, fibril formation, and neuronal loss in MPTP and 6-hydroxydopamine (6-OHDA) PD models (Yang et al., 2014; Wang et al., 2017, 2018). Curcumin-loaded liposomes functionalized with N-trimethyl chitosan chloride and D-α-tocopheryl PEG 1000 succinate decreased membrane dripping and enhanced loading efficiency (Chen et al., 2012; Wang et al., 2018). Resveratrol-loaded fusogenic liposomes profoundly decreased intracellular ROS generation and apoptosis in aged cells (Csiszar et al., 2015; Wang et al., 2018). Liposomal resveratrol has been found to protect from abnormal motor behavior and nigral cell loss (Wang et al., 2011; Wang et al., 2018).

4.4 Fibroblast Growth Factor

Fibroblast growth factor-20 (FGF-20), a paracrine candidate of FGF family specifically expressed in substantia nigra pars compacta, plays an essential role in protecting dopaminergic neurons (Beenken and Mohammadi, 2009; Ohmachi et al., 2000; Thuret et al., 2004; Niu et al., 2018). Nevertheless, low solubility of the bacterial recombinant human FGF-20 (rhFGF20) and inability to span BBB impeded its clinical use. To overcome these potential challenges, focused ultrasound guided, liposomal formulation with small ubiquitin-related modifier (SUMO)—tethered rhFGF20 has been designed, which showed enhanced BBB transport and protected dopaminergic neurons in 6-OHDA-lesioned PD rat model (Niu et al., 2018).

4.5 Solid lipid nanoparticles

SLNs with small size, biodegradability and biocompatibility adequate loading efficacy, and lipid nature have been considered the best nanostructures to accomplish brain spanning (Dal Magro et al., 2017, Khan et al., 2018). SLNs have solid lipid core and low melting point at physiological temperature (Dal Magro et al., 2017, Khan et al., 2018). Ease of surface functionalization of SLNs with targeting moieties to circumvent barriers such as BBB made them the most amenable DDS in the treatment of neurological disorders (Dal Magro et al., 2017, Khan et al., 2018). The solid lipid matrix facilitates the stability and slow release of loaded drugs (Cacciatore et al., 2016, Siddiq et al., 2018). They offer reproducibility by using several strategies and increased scale-up possibility (Roney et al., 2005, Siddiq et al., 2018). Therefore, they are considered as the substantial delivery systems (Sozio et al., 2014; Gastaldi et al., 2014, Siddiq et al., 2018). Surface functionalization of SLNs with hydrophilic polymers or surfactants has been found to avert phagocytosis (Alexis et al., 2008, Siddiq et al., 2018). Although amenable characteristics of SLNs made them substantial DDS in the treatment of NDs, yet low drug loading capacity limits their success (Khan et al., 2018). In addition, the aggregation of SLNs in reticuloendothelial system impedes their brain-targeted delivery (Siddiq et al., 2018).

Another type of lipid-based nanoformulation is nanostructured lipid carriers (NLCs), in which liquid lipid matrix is present and plays an important role in nanotechnological drug delivery (Khan et al., 2018).

Surface functionalization of NLCs with cationic surfactants enhanced their brain ferrying ability significantly (Scioli Montoto et al., 2018, Khan et al., 2018). Serine-based surfactants have been found to exhibit higher stability and increased transport through cell membranes and BBB (Mendes et al., 2018; Khan et al., 2018).

5. Demerits of liposomes

Although liposomes are valuable DDS in brain targeting, yet a few challenges such as unstable nature and vulnerability to break down during therapeutic delivery impede their success (Wang et al., 2018). To overcome these issues, their surface is functionalized with polysaccharides, polymers, peptides, aptamers, or antibodies, and brain-specific targeting was accomplished (Dua et al., 2012; Raposo and Stoorvogel, 2013; Kreuter, 2014; Eloy et al., 2017; Wang et al., 2018). Despite the adequate stability of liposomal curcumin in diverse buffers, membrane dripping of loaded therapeutics negatively influences bioavailability and therapy (Allijn et al., 2016, Wang et al., 2018). The important limitation of cationic liposomes is nonspecific delivery of therapeutic and the requirement of high dosage.

6. Conclusions and future perspectives

Liposomes have been considered the most appropriate delivery systems for the brain, although several types of systems are in use currently. Manifold drugs and natural compounds were delivered using liposomes. The substantial hitherto developed liposomal formulations adequately enhanced drug loading, controlled release of the loaded therapeutic for long duration, and extended systemic circulation profoundly. Nevertheless, a few demerits in some formulations such as drug dripping limit their success. Therefore, there is indeed a great need to develop more robust formulations to overcome these challenges.

References

Abou el Ela, A.E.S.F., El Khatib, M.M., Salem-Bekhit, M.M., 2017. Design, Characterization and microbiological evaluation of microemulsion based gel of griseofulvin for topical delivery system. Biointerface Res. Appl. Chem. 7, 2277–2285.

Abbina, S., Parambath, A., 2018. Chapter 14—pegylation and its alternatives: a summary. In: Parambath, A. (Ed.), Engineering of Biomaterials for Drug Delivery Systems. Woodhead Publishing, Sawston, UK, pp. 363–376.

Agid, Y., Destée, A., Durif, F., Montastruc, J.L., Pollak, P., 1997. Tolcapone, bromocriptine, and Parkinson's disease. Lancet 350, 712–713.

Agrawal, M., Tripathi Ajazuddin, D.K., Saraf, S., Saraf, S., Antimisiaris, S.G., Mourtas, S., et al., 2017. Recent advancements in liposomes targeting strategies to cross blood-brain barrier (BBB) for the treatment of Alzheimer's disease. J. Control. Release 260, 61–77.

Aharon, D., Weitman, H., Ehrenberg, B., 2011. The effect of liposomes' surface electric potential on the uptake of hematoporphyrin. Biochim. Biophys. Acta Biomembr. 1808, 2031–2035.

Alexis, F., Pridgen, E., Molnar, L.K., Farokhzad, O.C., 2008. Factors affecting the clearance and biodistribution of polymeric nanoparticles. Mol. Pharm. 5, 505–515.

Allijn, I.E., Schiffelers, R.M., Storm, G., 2016. Comparison of pharmaceutical nanoformulations for curcumin: enhancement of aqueous solubility and carrier retention. Int. J. Pharm. 506, 407–413.

Ay, M., Luo, J., Langley, M., Jin, H., Anantharam, V., Kanthasamy, A., et al., 2017. Molecular mechanisms underlying protective effects of quercetin against mitochondrial dysfunction and progressive dopaminergic neurodegeneration in cell culture and MitoPark transgenic mouse models of Parkinson's Disease. J. Neurochem. 141, 766–782.

Ballabh, P., Braun, A., Nedergaard, M., 2004. The blood-brain barrier: an overview: structure, regulation, and clinical implications. Neurobiol Dis 16, 1–13.

Beenken, A., Mohammadi, M., 2009. The FGF family: biology, pathophysiology and therapy. Nat. Rev. Drug Discov. 8, 235–253.

Cacciatore, I., Ciulla, M., Fornasari, E., Marinelli, L., Di Stefano, A., 2016. Solid lipid nanoparticles as a drug delivery system for the treatment of neurodegenerative diseases. Expert Opin. Drug Deliv. 13, 1121–1131.

Campos-Martorell, M., Cano-Sarabia, M., Simats, A., Hernandez-Guillamon, M., Rosell, A., Maspoch, D., et al., 2016. Charge effect of a liposomal delivery system encapsulating simvastatin to treat experimental ischemic stroke in rats. Int. J. Nanomed. 11, 3035–3048.

Cao, X., Hou, D., Wang, L., Li, S., Sun, S., Ping, Q., et al., 2016. Effects and molecular mechanism of chitosan-coated levodopa nanoliposomes on behavior of dyskinesia rats. Biol. Res. 49, 32.

Cetin, M., Aytekin, E., Yavuz, B., Bozdag-Pehlivan, S., 2017. Chapter 7—nanoscience in targeted brain drug delivery. In: Gursoy-Ozdemir, Y., Bozdag-Pehlivan, S., Sekerdag, E. (Eds.), Nanotechnology Methods for Neurological Diseases and Brain Tumors. Academic Press, Cambridge, MA, USA, pp. 117–138.

Chen, H., Wu, J., Sun, M., Guo, C., Yu, A., Cao, F., et al., 2012. N-trimethyl chitosan chloride-coated liposomes for the oral delivery of curcumin. J. Liposome Res. 22, 100–109.

Chen, Z.L., Huang, M., Wang, X.R., Fu, J., Han, M., Shen, Y.Q., et al., 2016a. Transferrin-modified liposome promotes α-mangostin to penetrate the blood-brain barrier. Nanomed. Nanotechnol. Biol. Med. 12, 421–430, 2016.

Chen, W., Li, H., Liu, Z., Yuan, W., 2016b. Lipopolyplex for therapeutic gene delivery and its application for the treatment of Parkinson's disease. Front. Aging Neurosci. 8, 68.

Cole, G.M., Teter, B., Frautschy, S.A., 2007. Neuroprotective effects of curcumin. Adv. Exp. Med. Biol. 595, 197–212.

Csiszar, A., Csiszar, A., Pinto, J.T., Gautam, T., Kleusch, C., Hoffmann, B., et al., 2015. Resveratrol encapsulated in novel fusogenic liposomes activates Nrf2 and attenuates oxidative stress in cerebromicrovascular endothelial cells from aged rats. J. Gerontol.: Biol Sci Med Sci 70, 303–313.

Dal Magro, R., Ornaghi, F., Cambianica, I., Beretta, S., Re, F., Musicanti, C., et al., 2017. ApoE-modified solid lipid nanoparticles: a feasible strategy to cross the blood-brain barrier. J. Control. Release 249, 103–110.

Darvesh, A.S., Carroll, R.T., Bishayee, A., Novotny, N.A., Geldenhuys, W.J., Van der Schyf, C.J., 2012. Curcumin and neurodegenerative diseases: a perspective. Expert Opin. Investig. Drugs 21, 1123–1140.

Dass, C.R., Choong, P.F., 2006. Targeting of small molecule anticancer drugs to the tumour and its vasculature using cationic liposomes: lessons from gene therapy. Cancer Cell Int. 6, 17.

Davoodi, S.D., Saghavaz, B.H., 2017. Optimal synthesis and characterization of magnetic CuMnFe2O4 nanoparticles coated by peg for drug delivery. Biointerface Res. Appl. Chem. 7, 2249–2252.

Di Stefano, A., Carafa, M., Sozio, P., Pinnen, F., Braghiroli, D., Orlando, G., et al., 2004. Evaluation of rat striatal L-dopa and DA concentration after intraperitoneal administration of L-dopa prodrugs in liposomal formulations. J. Control. Release 99, 293–300.

Di Stefano, A., Sozio, P., Iannitelli, A., Marianecci, C., Santucci, E., Carafa, M., 2006. Maleic- and fumaric-diamides of (O,O-diacetyl)-L-Dopa-methylester as anti-Parkinson prodrugs in liposomal formulation. J. Drug Target. 14, 652–661.

Dong, X., 2018. Current strategies for brain drug delivery. Theranostics 8, 1481–1493.

Dua, J., Rana, A.C., Bhandari, A.K., 2012. Liposome: methods of preparation and applications. Int J Pharm Stud Res 3, 14–20.

Eloy, J.O., Petrilli, R., Trevizan, L.N.F., Chorilli, M., 2017. Immunoliposomes: a review on functionalization strategies and targets for drug delivery. Colloids Surf. B Biointerfaces 159, 454–467.

Fonseca-Santos, B., Gremiao, M.P.D., Chorilli, M., 2015. Nanotechnology-based drug delivery systems for the treatment of Alzheimer's disease. Int. J. Nanomed. 10, 4981–5003.

Gastaldi, L., Battaglia, L., Peira, E., Chirio, D., Muntoni, E., Solazzi, I., et al., 2014. Solid lipid nanoparticles as vehicles of drugs to the brain: current state of the art. Eur. J. Pharm. Biopharm. 87, 433–444.

Gregoriadis, G., 2008. Liposome research in drug delivery: the early daysJ. Drug Target 16, 520–524.

Gurturk, Z., Tezcaner, A., Dalgic, A.D., Korkmaz, S., Keskin, D., 2017. Maltodextrin modified liposomes for drug delivery through the blood-brain barrier. Medchemcomm 8, 1337–1345.

Hu, Y., Rip, J., Gaillard, P.J., de Lange, E.C.M., Hammarlund-Udenaes, M., 2017. The impact of liposomal formulations on the release and brain delivery of methotrexate: an in vivo microdialysis study. J. Pharm. Sci. 106, 2606–2613.

Jain, N.K., Rana, A.C., Jain, S.K., 1998. Brain drug delivery system bearing dopamine hydrochloride for effective management of Parkinsonism. Drug Dev. Ind. Pharm. 24, 671–675.

Karmali, P.P., Chaudhuri, A., 2007. Cationic liposomes as non-viral carriers of gene medicines: resolved issues, open questions, and future promises. Med. Res. Rev. 27, 696–722.

Kang, Y.J., Cutler, E.G., Cho, H., 2018. Therapeutic nanoplatforms and delivery strategies for neurological disorders. Nano Converg 5, 35.

Kelkar, S.S., Reineke, T.M., 2011. Theranostics: combining imaging and therapy. Bioconjug. Chem. 22, 1879–1903.

Khan, A.R., Yang, X., Fu, M., Zhai, G., 2018. Recent progress of drug nanoformulations targeting to brain. J. Control. Release 291, 37–64.

Kreuter, J., 2014. Drug delivery to the central nervous system by polymeric nanoparticles: what do we know? Adv. Drug Deliv. Rev. 71, 2–14.

Kristensen, M., Brodin, B., 2017. Routes for drug translocation across the blood-brain barrier: exploiting peptides as delivery vectors. J. Pharm. Sci. 106, 2326–2334.

Lai, F., Fadda, A.M., Sinico, C., 2013. Liposomes for brain delivery. Expert Opin. Drug Deliv. 10, 1003–1022.

Lakkadwala, S., Singh, J., 2018. Dual functionalized 5-fluorouracil liposomes as highly efficient nanomedicine for glioblastoma treatment as assessed in an in vitro brain tumor model. J. Pharm. Sci. 107, 2902–2913.

Lakkadwala, S., Singh, J., 2019. Co-delivery of doxorubicin and erlotinib through liposomal nanoparticles for glioblastoma tumor regression using an in vitro brain tumor model. Colloids Surf. B Biointerfaces 173, 27–35.

Lockman, P.R., Koziara, J.M., Mumper, R.J., Allen, D.D., 2004. Nanoparticle surface charges alter blood–brain barrier integrity and permeability. J. Drug Target. 12, 635–641.

Lopalco, A., Cutrignelli, A., Denora, N., Lopedota, A., Franco, M., Laquintana, V., et al., 2018. Transferrin functionalized liposomes loading dopamine HCl: development and permeability studies across an in vitro model of human blood–brain barrier. Nanomaterials 8.

Loureiro, J.A., Gomes, B., Fricker, G., Cardoso, I., Ribeiro, C.A., Gaiteiro, C., et al., 2015. Dual ligand immunoliposomes for drug delivery to the brain. Colloids Surf. B Biointerfaces 134, 213–219.

Md, S., Haque, S., Sahni, J.K., Baboota, S., Ali, J., 2011. New non-oral drug delivery systems for Parkinson's disease treatment. Expert Opin. Drug Deliv. 8, 359–374.

Mendes, M., Miranda, A., Cova, T., Goncalves, L., Almeida, A.J., Sousa, J.J., et al., 2018. Modeling of ultra-small lipid nanoparticle surface charge for targeting glioblastoma. Eur. J. Pharm. Sci. 117, 255–269.

Moretti, R., Torre, P., Vilotti, C., Antonello, R.M., Pizzolato, G., 2007. Rivastigmine and Parkinson dementia complex. Expert Opin. Pharmacother. 8, 817–829.

Nair, M.P.N., Saiyed, S.M., September 1, 2011. Inventors; the Florida International University Board of Trustees, Assignee. Magnetic Nanodelivery of Therapeutic Agents across the Blood Brain Barrier. United States Patent US 20110213193 A1.

Nisi Zhang, F.Y., 2018. Localized delivery of curcumin into brain with polysorbate 80-modified cerasomes by ultrasound-targeted microbubble destruction for improved Parkinson's disease therapy. Theranostics 8, 2264–2276.

Niu, J., Xie, J., Guo, K., Zhang, X., Xia, F., Zhao, X., et al., 2018. Efficient treatment of Parkinson's disease using ultrasonography-guided rhFGF20 proteoliposomes. Drug Deliv. 25, 1560–1569.

Nutt, J.G., Carter, J.H., Sexton, G.J., 2004. The dopamine transporter: importance in Parkinson's disease. Ann. Neurol. 55, 766–773.

Obulesu, M., 2019. Chapter 5: blood brain barrier targeted nanotechnological advances. In: Alzheimer's Disease Theranostics. Elsevier, pp. 25–31.

Obulesu, M., Jhansilakshmi, M., 2016. Chapter 12: liposomes in Apoptosis induction and cancer therapy. In: Apoptosis Methods in Toxicology. Springer publishers, pp. 231–237.

Ohmachi, S., Watanabe, Y., Mikami, T., Kusu, N., Ibi, T., Akaike, A., et al., 2000. FGF-20, a novel neurotrophic factor, preferentially expressed in the substantia nigra pars compacta of rat brain. Biochem. Biophys. Res. Commun. 277, 355–360.

Peng, Y., Zhao, Y., Chen, Y., Yang, Z., Zhang, L., Xiao, W., et al., 2018. Dual-targeting for brain-specific liposomes drug delivery system: synthesis and preliminary evaluation. Bioorg. Med. Chem. 26, 4677–4686.

Pignatello, R., Fuochi, V., Petronio, G.P., Greco, A.S., Furneri, P.M., 2017. Formulation and characterization of erythromycin–loaded solid lipid nanoparticles. Biointerface Res Appl Chem 7, 2145–2150.

Poewe, W., Seppi, K., Tanner, C.M., Halliday, G.M., Brundin, P., Volkmann, J., et al., 2017. Parkinson disease. Nat Rev Dis Primers 3, 17013.

Qu, M., Lin, Q., He, S., Wang, L., Fu, Y., Zhang, Z., et al., 2018a. A brain targeting functionalized liposomes of the dopamine derivative N-3,4-bis(pivaloyloxy)-dopamine for treatment of Parkinson's disease. J. Control. Release 277, 173–182.

Qu, M., Lin, Q., He, S., Wang, L., Fu, Y., Zhang, Z., et al., 2018b. A brain targeting functionalized liposomes of the dopamine derivative N-3,4-bis(pivaloyloxy)-dopamine for treatment of Parkinson's disease. J. Control. Release 277, 173–182.

Ramos-Cabrer, P., Campos, F., 2013. Liposomes and nanotechnology in drug development: focus on neurological targets. Int. J. Nanomed. 8, 951–960.

Raposo, G., Stoorvogel, W., 2013. Extracellular vesicles: exosomes, microvesicles, and friends. J. Cell Biol. 200, 373–383.

Roney, C., Kulkarni, P., Arora, V., Antich, P., Bonte, F., Wu, A., et al., 2005. Targeted nanoparticles for delivery through blood brain barrier for Alzheimer's disease. J. Control. Release 108, 193–214.

Scioli Montoto, S., Sbaraglini, M.L., Talevi, A., Couyoupetrou, M., Di Ianni, M., Pesce, G.O., et al., 2018. Carbamazepine-loaded solid lipid nanoparticles and nanostructured lipid carriers: physicochemical characterization and in vitro/in vivo evaluation. Colloids Surf. B Biointerfaces 167, 73–81.

Senior, J.H., 1987. Fate and behavior of liposomes in vivo: a review of controlling factors. Crit. Rev. Ther. Drug Carrier Syst. 3, 123–193.

Shrestha, J.P., Subedi, Y.P., Chen, L., Chang, C.W.T., 2015. A mode of action study of cationic anthraquinone analogs: a new class of highly potent anticancer agents. Med. Chem. Commun. 6, 2012–2022.

Sian, D., Dexter, T., Lees, A.J., Daniel, S., Agid, Y., Javoy-Agid, F., et al., 1994. Alterations in glutathione levels in Parkinson's disease and other neurodegenerative disorders affecting basal ganglia. Ann. Neurol. 36, 348–355.

Siddiqi, K.S., Husen, A., Sohrab, S.S., Yassin, M.O., 2018. Recent Status of Nanomaterial Fabrication and Their Potential Applications in Neurological Disease Management. Nanoscale Res Lett 13, 231.

Sies, H., 1999. Glutathione and its role in cellular functions. Biol. Med. 27, 916–921.

Sozio, P., Fiorito, J., Di Giacomo, V., Di Stefano, A., Marinelli, L., Cacciatore, I., et al., 2014. Haloperidol metabolite II prodrug: asymmetric synthesis and biological evaluation on rat C6 glioma cells. Eur. J. Med. Chem. 90, 1–9.

Spillantini, M.G., Schmidt, M.L., Lee, V.M., Trojanowski, J.Q., Jakes, R., Goedert, M., 1997. α-Synuclein in Lewy bodies. Nature 388, 839.

Spuch, C., Navarro, C., 2011. Liposomes for targeted delivery of active agents against neurodegenerative diseases (Alzheimer's disease and Parkinson's disease). J. Drug Deliv. 469679.

Teleanu, D.M., Chircov, C., Grumezescu, A.M., Volceanov, A., Teleanu, R.I., 2018. Blood-brain delivery methods using nanotechnology. Pharmaceutics 10.

Thanvi, B., Lo, N., Robinson, T., 2007. Levodopa-induced dyskinesia in Parkinson's disease: clinical features, pathogenesis, prevention and treatment. Postgrad. Med. J. 83, 384–388.

Thuret, S., Bhatt, L., O'Leary, D.D., Simon, H.H., 2004. Identification and developmental analysis of genes expressed by dopaminergic neurons of the substantia nigra pars compacta. Mol. Cell. Neurosci. 25, 394–405.

Vieira, D.B., Gamarra, L.F., 2016. Getting into the brain: liposome-based strategies for effective drug delivery across the blood-brain barrier. Int. J. Nanomed. 11, 5381–5414.

Wakaskar, R.R., 2017. Promising effects of nanomedicine in cancer drug delivery. J. Drug Target. 26, 319–324.

Wang, Y., Xu, H., Fu, Q., Ma, R., Xiang, J., 2011. Protective effect of resveratrol derived from Polygonum cuspidatum and its liposomal form on nigral cells in parkinsonian rats. J. Neurol. Sci. 304, 29–34.

Wang, Z.Y., Liu, J.Y., Yang, C.B., Malampati, S., Huang, Y.Y., Li, M.X., et al., 2017. Neuroprotective natural products for the treatment of Parkinson's disease by targeting the autophagy-lysosome pathway: a systematic review. Phytother Res. 31, 1119–1127.

Wang, Z.Y., Sreenivasmurthy, S.G., Song, J.X., Liu, J.Y., Li, M., 2018. Strategies for brain-targeting liposomal delivery of small hydrophobic molecules in the treatment of neurodegenerative diseases. Drug Discov. Today S1359–6446, 30161–30162.

Wen, C.J., Zhang, L.W., Al-Suwayeh, S.A., Yen, T.C., Fang, J.Y., 2012. Theranostic liposomes loaded with quantum dots and apomorphine for brain targeting and bioimaging. Int. J. Nanomed. 7, 1599–1611.

Xiang, Y., Wu, Q., Liang, L., Wang, X., Wang, J., Zhang, X., et al., 2012a. Chlorotoxin-modified stealth liposomes encapsulating levodopa for the targeting delivery against the Parkinson's disease in the MPTP-induced mice model. J. Drug Target. 20, 67–75.

Xiang, Y., Wu, Q., Liang, L., Wang, X., Wang, J., Zhang, X., et al., 2012b. Chlorotoxin-modified stealth liposomes encapsulating levodopa for the targeting delivery against the Parkinson's disease in the MPTP-induced mice model. J. Drug Target. 20, 67–75.

Yang, J., Song, S., Li, J., Liang, T., 2014. Neuroprotective effect of curcumin on hippocampal injury in 6- OHDA-induced Parkinson's disease rat. Pathol. Res. Pract. 210, 357–362.

Yurasov, V.V., Podgornyi, G.N., Kucheryanu, V.G., Kudrin, V.S., Nikushkin, E.V., Zhigal'tsev, I.V., et al., 1996. Effects of L-Dopa-carrying liposomes on striatal concentration of dopamine and its metabolites and phospholipid metabolism in experimental Parkinson's syndrome. Bull. Exp. Biol. Med. 122, 1180–1183.

Yurasov, V.V., Kucheryanu, V.G., Kudrin, V.S., Zhigal'tsev, I.V., Nikushkin, E.V., Sandalov, Yu, G., et al., 1997. Effect of long-term parenteral administration of empty and L-Dopa-loaded liposomes on the turnover of dopamine and its metabolites in the striatum of mice with experimental Parkinson's syndrome. Bull. Exp. Biol. Med. 123, 126–129.

Zhang, N., Yan, F., Liang, X., Wu, M., Shen, Y., Chen, M., et al., 2018. Localized delivery of curcumin into brain with polysorbate 80-modified cerasomes by ultrasound-targeted microbubble destruction for improved Parkinson's disease therapy. Theranostics 8, 2264–2277.

Zlokovic, B.V., 2008. The blood–brain barrier in health and chronic neurodegenerative disorders. Neuron 57, 178–201.

Further reading

Obulesu, M., 2018. Chapter: 16. Multifarious therapeutic avenues for Alzheimer's disease. In: Pathology, Prevention and Therapeutics of Neurodegenerative Disorders. Springer publishers, pp. 185–190.

CHAPTER 8

Nasal delivery nanoparticles: An insight into novel Parkinson's disease therapeutics

1. Introduction

Parkinson's disease (PD) is one among the neurodegenerative disorders which affects 1%–1.5% people aged more than 60 years and 3% people aged more than 65 years (Alves et al., 2008; Bi et al., 2016; Hao et al., 2017; Yan et al., 2018). The prevalence of PD has also been observed to grow with the aging population and enhanced life expectancy (Leyva-Gomez et al., 2015; Bi et al., 2016). A few antiparkinsonian drugs such as dopamine agonists (apomorphine), monoamine oxidase type B inhibitors, and cholinesterase inhibitors (donepezil) induce adverse side effects despite the partial symptomatic relief in PD patients (Di Stefano et al., 2009; Kulkarni et al., 2015; Bi et al., 2016). Although oral (Ghaffari et al., 2018), intraperitoneal (Chen et al., 2018), intravenous (Herring et al., 2017), intracerebral (Yan et al., 2018), transdermal (Ogawa et al., 2018), and intranasal drug delivery have been used to overcome PD till date, intranasal delivery has been found to be the most promising (see Table 8.1).

2. Nasal delivery

Nasal drug delivery has been confined to local problems such as nasal infections, rhinorrhea, nasal congestion, allergic, or chronic rhinosinusitis (Pozzoli et al., 2016; Sonvico et al., 2018). However, more recently, it has been used for complicated diseases such as Alzheimer's disease (AD) and PD (Sonvico et al., 2018). Blood–brain barrier (BBB) with its microvascular endothelial cells and tight junctions mediates intra- and intercellular signaling pathways, in turn playing an essential role in central nervous system (CNS) homeostasis (Neuwelt et al., 2011; Bi et al., 2016).

Parkinson's Disease Therapeutics
ISBN 978-0-12-819882-7
https://doi.org/10.1016/B978-0-12-819882-7.00008-8

Table 8.1 List of brain-targeted drugs and nanoparticles (NPs) through nasal delivery.

Therapeutic	Materials	Result	Reference
Fluorescein-labeled 4 kDa dextran	Sodium hyaluronate	Enhanced the brain localization of dextran	Horvat et al. (2009), Sonvico et al. (2018)
Rivastigmine	Electrosteric stealth liposomes	Enhanced blood–brain barrier translocation and therapeutic efficacy	Nageeb El-Helaly et al., 2017, Wang et al. (2018)
Fexofenadine	Chitosan (CS)-conjugated liposomes	Reduced drug dripping	Qiang et al. (2012), Wang et al. (2018)
Rotigotine	Lactoferrin-conjugated PEG-PLGA NPs	Improved therapeutic efficacy	Bi et al. (2016)
CS NPs			
Rotigotine	CS NPs	Substantial nose to brain delivery	Tzeyung et al. (2019)
Pramipexole dihydrochloride		Increased therapeutic efficacy	Raj et al. (2017)
Selegiline		12-fold enhanced selegiline concentrations in the brain and plasma	Sridhar et al. (2018)
Bromocriptine		Increased brain delivery	Md et al. (2013)
Ropinirole			Jafarieh et al. (2015)
Rasagiline			Mittal et al. (2014)
Pramipexole			Raj et al. (2017), Sonvico et al. (2018)
Dopamine	Glycol CS	Enhance dopamine encapsulation	Sonvico et al. (2018)

Unfortunately, it also impedes the therapeutic delivery to the brain (Chapter 3, Bi et al., 2016). To circumvent this biological riddle, nose to brain delivery of drugs through olfactory and trigeminal nerve pathways by evading BBB has been in progress currently (Ali et al., 2010; Lalani et al., 2014; Kulkarni et al., 2015; Bi et al., 2016). More specifically, trigeminal nerve facilitates the entry of drugs to the posterior region of the brain

(Thorne et al., 2004; Chapman et al., 2013; Sonvico et al., 2018). As olfactory neuroepithelium is the only section that is unprotected by BBB and can feasibly be communicated with the external environment, this has been treated the most feasible access to the brain (Pardeshi and Belgamwar, 2013; Sonvico et al., 2018). Intranasal delivery has also been considered to be more appropriate compared to intravenous route as it overcomes BBB in therapeutic delivery (Grassin-Delyle et al., 2012; Tzeyung et al., 2019). There is also a possibility of the therapeutic delivery through systemic circulation which may occur indirectly after intranasal introduction (Quintana et al., 2016; Feng et al., 2018; Wang et al., 2018; Teleanu et al., 2018; Mo et al., 2019).

Nasal delivery confers substantial advantages which include decreased drug delivery to nontarget sites, evading hepatic first pass metabolism and feasibility for drugs such as rasagiline, alginate, tarenflurbil, and piperine (Haque et al., 2014; Elnaggar et al., 2015; Muntimadugu et al., 2016; Mittal et al., 2016; Yan et al., 2018). In line with this, sodium hyaluronate enhanced the brain localization of a high molecular weight hydrophilic compound fluorescein-labeled 4 kDa dextran after intranasal introduction into rats (Horvat et al., 2009; Sonvico et al., 2018).

3. Challenges in nasal delivery

Challenges associated with nasal delivery of drugs include encountering nasal metabolizing enzymes and mucociliary elimination (Wang et al., 2018). In addition, characteristics of small molecules such as size and lipophilicity influence their delivery to CNS after intranasal introduction (Agrawal et al., 2018; Wang et al., 2018). To overcome these challenges, wealth of studies developed potential nasal drug carriers such as nanoparticles (NPs), liposomes, microemulsions and microspheres, nanocrystals, nanostructured lipid carriers (NLCs), albumin NPs, gelatin NPs, and dendrimers (Tafaghodi et al., 2006; Heurtault et al., 2010; Wang et al., 2012; Kurti et al., 2013; Vaka et al., 2013; Win-Shwe et al., 2014; Alam et al., 2014; Bartos et al., 2015; Khan et al., 2016; Bi et al., 2016; Wong et al., 2018; Sonvico et al., 2018).

3.1 Nanotechnology

NPs veil the physicochemical characteristics of drugs and offer advantages such as controlled drug release, lessened drug toxicity, ameliorated bioavailability, biodistribution, and therapeutic potential (Ali et al., 2010;

Gomes et al., 2014; Bi et al., 2016). Polymeric NPs facilitated nasal delivery and enhanced therapeutic efficacy of loaded therapeutics because of their small size, feasible localization into cells, and ability to aggregate drugs at the target sites (Soppimath et al., 2001; Kreuter, 2001; Raj et al., 2017). Numerous polymers which enhance nasal drug delivery include natural (gums, gelatin, alginates, and starch), semisynthetic (cellulose derivatives such as hydroxypropyl, methyl, hydroxypropylmethyl, and carboxymethyl cellulose), and synthetic (crospovidone, polyacrylates, and poly-methacrylates) polymers (Ugwoke et al., 2005; Sonvico et al., 2018).

3.2 Liposomes

Electrosteric stealth (ESS) liposomes loaded with rivastigmine showed enhanced BBB translocation and therapeutic efficacy compared to rivastigmine alone when introduced through nasal route (Nageeb El-Helaly et al., 2017; Wang et al., 2018). In addition, in vivo pharmacokinetic study has also proved 486% of bioavailability of ESS liposome-loaded rivastigmine compared to rivastigmine alone (Nageeb El-Helaly et al., 2017; Wang et al., 2018). In another study, liposomes were designed using cholesterol, poly(ethylene) glycol (PEG), and 1,2-distearoyl-sn-glycero-3-phosphocholine (DSPC). Introduction of liposomal donepezil through intranasal route remarkably improved bioavailability of donepezil (Al Asmari et al., 2016; Wang et al., 2018). Chitosan (CS)-conjugated liposomes have been used to accomplish stability even at 4°C for 6 months, reduced drug dripping by 10%, nasal delivery of fexofenadine (Qiang et al., 2012; Wang et al., 2018). The stability of liposomes can also be improved by adjusting factors such as temperature and cholesterol inclusion in the lipid bilayer (Sukowski et al., 2005; Wang et al., 2018).

3.3 Surface functionalization

As PEG on NPs surface suppresses cell surface communication, NPs functionalization with biological ligands is more amenable option to accomplish nose to brain delivery (Verma and Stellacci, 2010; Bi et al., 2016). Lactoferrin (Lf), an iron-binding glycoprotein of the transferrin family with a molecular weight of 80 kDa, has been found to be overexpressed in the capillaries and neurons linked to age-associated neurodegenerative disorders such as AD, PD, and amyotrophic lateral sclerosis (Ward et al., 2002; Bi et al., 2016). Accordingly, Lf exhibits enhanced brain translocation compared to transferrin (Qian and Wang., 1998, Ji et al., 2006, Bi et al., 2016). Hence, Lf enhances nose to brain delivery of NPs (Bi et al., 2016). PEG—poly(lactic-co-glycolic acid) (PEG-PLGA) NPs surface-functionalized with Lf showed considerably enhanced rotigotine concentration in the striatum region of brain after intranasal administration and improved therapeutic efficacy (Bi et al., 2016).

Mistry et al., (2015) evaluated the permeation of carboxylate-modified fluorescent polystyrene NPs measuring 20,100, and 200 nm in size (ζ potential: approximately −42 mV) with surface-functionalized NPs designed by tethering with CS (48, 163, or 276 nm; ζ potential approximately +30 mV) or polysorbate 80 (ζ potential approximately −21 mV) in porcine olfactory epithelium. Surprisingly, except polysorbate 80—tethered (PEGylated) particles, none of the particles spanned the nasal epithelium after 90 min (Sonvico et al., 2018).

3.4 Chitosan Nanoparticles

CS being the deacetylated form of chitin has long been widely used in the synthesis and design of NPs (Raj et al., 2017). Its potential to be biocompatible, biodegradable, and mucoadhesive in nature makes it an amenable material in the NP preparation (Nagpal et al., 2010; Di Gioia et al., 2015; Raj et al., 2017). In addition to the abovementioned benefits, CS also ameliorates the permeability through tight junctions between epithelial cells in mucosal tissues (Deli, 2009; Sonvico et al., 2018). Plethora of nasal formulations with capability to enhance viscosity and/or offer bioadhesion such as hydrophilic polymers, to encounter mucociliary clearance, increase formulation residence time, ameliorate systemic bioavailability, and decrease nasal absorption difference (Marttin et al., 1998; Sonvico et al., 2018). Interestingly, mucoadhesive nature of CS plays a pivotal role in nasal delivery because nasal cavity is encountered with the problem of constant ciliary clearance (Raj et al., 2017). Several lines of evidence have shown the ionic interaction between positive and negative charges of amino groups of CS and sialic acid groups of mucin, respectively (Henriksen et al., 1996; Ozsoy et al., 2009; Raj et al., 2017). In line with this, CS has been found to enhance the retention time of therapeutics in the nasal cavity, in turn facilitating the enhanced drug distribution through nasal epithelium (Illum, 2002; Raj et al., 2017). In addition, CS and its derivatives substantially ameliorate nasal absorption (Casettari and Illum, 2014; Raj et al., 2017).

Rotigotine-encapsulated CS NPs have been designed which showed substantial nose to brain delivery of the therapeutic (Tzeyung et al., 2019). Despite the remarkably high therapeutic efficacy of rotigotine, decreased oral bioavailability, low plasma half-life of 5—7 h, and inability to span BBB impede its success (Md et al., 2011; Li et al., 2014; Tzeyung et al., 2019). Studies also reported remarkably high therapeutic efficacy of pramipexole dihydrochloride—encapsulated CS NPs both in in vitro and in vivo (Raj et al., 2017). Selegiline, a vital antiparkinsonian agent, shows low oral bioavailability and increased toxicity (Sridhar et al., 2018). To overcome these issues, CS NPs have been designed, which showed 20- and 12-fold enhanced selegiline concentrations in the brain and plasma, respectively (Sridhar et al., 2018). Results also indicated ameliorated motor functions, considerably enhanced dopamine, catalase activity, and glutathione content in the brain (Sridhar et al., 2018). In nose to brain delivery, CS and low molecular weight pectins were found to enhance retention time of nasal formulations in the olfactory section of humans (Charlton et al., 2007; Sonvico et al., 2018). CS NPs developed by ionotropic gelation with tri-polyphosphate (TPP) have been used to deliver bromocriptine (Md et al., 2013), ropinirole (Jafarieh et al., 2015), rasagiline (Mittal et al., 2014), and pramipexole (Raj et al., 2017; Sonvico et al., 2018).

Despite the noteworthy merits of CS, a few limitations such as incomplete solubility at physiological pH and exhibition of positive charge only in acidic conditions hinder its success as a bioamenable material. However, to overcome these limitations, CS derivatives such as trimethyl CS have been extensively studied because of its consistent positive charge, potential water solubility, and capability to enhance penetration of the peptide through porcine nasal mucosa (Kumar et al., 2013; Sonvico et al., 2018). In line with this, glycol CS being water soluble form of CS has been utilized in the development of NPs by ionotropic gelation with TPP in association with sulfobutylether-β-cyclodextrin to enhance dopamine encapsulation (Sonvico et al., 2018). Although single intranasal introduction did not alter brain levels of neurotransmitter, yet repeated introduction showed profoundly improved dopamine levels in the ipsilateral striatum (Di Gioia et al., 2015; Sonvico et al., 2018).

3.5 Nanostructured lipid carriers

Gartziandia et al. (2016) have shown the influence of type of nanomaterial and surface charge in the transport of NPs through primary cell monolayers of rat olfactory mucosa. In this study, a fluorescent probe (DiR; 1-10-dioctadecyl-3,3,3′,3′-tetranethylindotricarbocyanine) was encapsulated to trace the translocation of NPs. Substantial differences were found based on the material type and surface charge. NLCs showed enhanced permeation compared to PLGA NPs with similar zeta potential (−23 mV). CS conjugation to NLCs was found to enhance permeation by threefold compared to unconjugated NLCs (Sonvico et al., 2018).

Gabal et al. (2014) studied the influence of surface charge on nasal delivery of NPs by designing anionic and cationic NLCs with similar size (175 and 160 nm, respectively) and positive and negative ζ potential (−34 and +34 mV, respectively). Interestingly, particles showed similar drug content and release kinetics in vitro (Sonvico et al., 2018). These NPs were loaded in a thermosensitive gel developed using poloxamers (407 and 188) and hydroxypropyl methylcellulose and introduced intranasally into albino rats. Both positive and negative NPs showed better efficacy, and no considerable difference between the particles was observed (Sonvico et al., 2018).

3.6 Peptide-based carriers

Kanazawa et al. (2015) studied the essential role of peptide-based nanocarriers in nasal delivery. In this study, an arginine-rich oligopeptide (with adhesive nature and carrying ability) was tethered with hydrophobic stearic acid or hydrophilic PEG—poly(ε-caprolactone) (PCL) block copolymer to develop two stable micellar formulations. To study biodistribution, an Alexa-dextran complex (molecular weight 10,000 Da) was used as a fluorescent probe. The stearate-peptide and PEG-PCL-peptide micelles showed 100 and 50 nm size and +20 and +15 mV ζ potential, respectively. Both carriers showed enhanced nasal and brain uptake compared to Alexa-dextran alone (Sonvico et al., 2018).

3.7 Polysaccharide-based Nanoparticles

Polysaccharides offer appreciable advantages in the synthesis of mucoadhesive nanocarriers. Accordingly, they show mucoadhesive characteristics, biomimetic mucosal recognition, biodegradability, biocompatibility, and feasibility to modify NPs chemically with targeting moieties (Sonvico et al., 2018). Polysaccharide-based NPs have been designed by adhering them to predesigned NPs, copolymerization, or covalent binding, resulting in surface functionalization or by developing polysaccharide-based NPs (Lemarchand et al., 2004; Sonvico et al., 2018).

4. Demerits of nose to brain delivery

Despite the worth considering therapeutic effects of nose to brain delivery, a few demerits such as toxicity to the nasal epithelium have been reported

(Sonvico et al., 2018). In line with this, positively charged CS conjugated NPs showed size-based increase in toxicity to porcine olfactory epithelium (Mistry et al., 2015; Sonvico et al., 2018). On the other hand, studies on animal models proved the potential safety of CS conjugated NPs (Casettari and Illum, 2014; Sonvico et al., 2018). Growing lines of evidence showed that pollutants and metal NPs induce toxic side effects after intranasal administration or inhalation (Simko and Mattsson, 2010; Lucchini et al., 2012; Karmakar et al., 2014; Sonvico et al., 2018).

5. Conclusions and future perspectives

Of the multifarious routes of drug administration, nasal delivery has been found to be the most promising route. The primary reasons for such enhanced efficacy include avoidance of first pass metabolism, bypassing BBB, and enhancing therapeutic loading in CNS (Yan et al., 2018). Conversely, a few nanomaterials showed toxicity to the olfactory epithelium. Therefore, there is a growing need to garner expertise from several fields such as medicine, chemistry, and nanotechnology to develop more robust therapeutic arsenals that do not show toxicity to the olfactory epithelium or other tissues.

References

Agrawal, M., Saraf, S., Saraf, S., Antimisiaris, S.G., Chougule, M.B., Shoyele, S.A., et al., 2018. Nose-to-brain drug delivery: an update on clinical challenges and progress towards approval of anti-alzheimer drugs. J. Control Release Off. J. Control Rel. Soc. 281, 139–177.

Alam, T., Pandit, J., Vohora, D., Aqil, M., Ali, A., Sultana, Y., 2014. Optimization of nanostructured lipid carriers of lamotrigine for brain delivery: in vitro characterization and in vivo efficacy in epilepsy. Expert Opin. Drug Deliv. 12, 181–194.

Ali, J., Ali, M., Baboota, S., Sahani, J.K., Ramassamy, C., Dao, L., et al., 2010. Potential of nanoparticulate drug delivery systems by intranasal administration. Curr. Pharmaceut. Des. 16, 1644–1653.

Al Asmari, A.K., Ullah, Z., Tariq, M., Fatani, A., 2016. Preparation, characterization, and *in vivo* evaluation of intranasally administered liposomal formulation of donepezil. Drug Des. Dev. Ther. 10, 205–215.

Alves, G., Forsaa, E.B., Pedersen, K.F., Dreetz Gjerstad, M., Larsen, J.P., 2008. Epidemiology of Parkinson's disease. J. Neurol. 255 (Suppl. 5), 18–32.

Bartos, C., Ambrus, R., Sipos, P., Budai-Szűcs, M., Csanyi, E., Gaspar, R., et al., 2015. Study of sodium hyaluronate-based intranasal formulations containing micro- or nanosized meloxicam particles. Int. J. Pharm. 491, 198–207.

Bi, C., Wang, A., Chu, Y., Liu, S., Mu, H., Liu, W., et al., 2016. Intranasal delivery of rotigotine to the brain with lactoferrin-modified PEG-PLGA nanoparticles for Parkinson's disease treatment. Int. J. Nanomed. 11, 6547–6559.

Casettari, L., Illum, L., 2014. Chitosan in nasal delivery systems for therapeutic drugs. J. Control. Release 190, 189–200.

Chen, Y., Hou, Y., Ge, R., Han, J., Xu, J., Chen, J., et al., 2018. Protective effect of roscovitine against rotenone-induced parkinsonism. Restor. Neurol. Neurosci. 36, 629–638.

Chapman, C.D., Frey, W.H., Craft, S., Danielyan, L., Hallschmid, M., Benedict, C., 2013. Intranasal treatment of central nervous system dysfunction in humans. Pharm. Res. 30, 2475–2484.

Charlton, S., Jones, N.S., Davis, S.S., Illum, L., 2007. Distribution and clearance of bioadhesive formulations from the olfactory region in man: effect of polymer type and nasal delivery device. Eur. J. Pharm. Sci. 30, 295–302.

Deli, M.A., 2009. Potential use of tight junction modulators to reversibly open membranous barriers and improve drug delivery. Biochim. Biophys. Acta 1788, 892–910.

Di Gioia, S., Trapani, A., Mandracchia, D., De Giglio, E., Cometa, S., Mangini, V., et al., 2015. Intranasal delivery of dopamine to the striatum using glycolchitosan/sulfobutylether -β-cyclodextrin based nanoparticles. Eur. J. Pharm. Biopharm. 94, 180–193.

Di Stefano, A., Sozio, P., Iannitelli, A., Cerasa, L.S., 2009. New drug delivery strategies for improved Parkinson's disease therapy. Expert Opin. Drug Deliv. 6, 389–404.

Elnaggar, Y.S., Etman, S.M., Abdelmonsif, D.A., Abdallah, O.Y., 2015. Intranasal piperine-loaded Chitosan nanoparticles as brain-targeted therapy in Alzheimer's disease: optimization, biological efficacy, and potential toxicity. J. Pharm. Sci. 104, 3544–3556.

Feng, Y., He, H., Li, F., Lu, Y., Qi, J., Wu, W., 2018. An update on the role of nanovehicles in nose-to-brain drug delivery. Drug Discov. Today 23, 1079–1088.

Gabal, Y.M., Kamel, A.O., Sammour, O.A., Elshafeey, A.H., 2014. Effect of surface charge on the brain delivery of nanostructured lipid carriers in situ gels via the nasal route. Int. J. Pharm. 473, 442–457.

Ghaffari, F., Hajizadeh Moghaddam, A., Zare, M., 2018. Neuroprotective effect of quercetin nanocrystal in a 6-hydroxydopamine model of Parkinson disease: biochemical and behavioral evidence. Basic Clin. Neurosci. 9, 317–324.

Gartziandia, O., Egusquiaguirre, S.P., Bianco, J., Pedraz, J.L., Igartua, M., Hernandez, R.M., et al., 2016. Nanoparticle transport across in vitro olfactory cell monolayers. Int. J. Pharm. 499, 81–89.

Gomes, M.J., Neves, J., Sarmento, B., 2014. Nanoparticle-based drug delivery to improve the efficacy of antiretroviral therapy in the central nervous system. Int. J. Nanomed. 9, 1757–1769.

Grassin-Delyle, S., Buenestado, A., Naline, E., Faisy, C., Blouquit-Laye, S., Couderc, L.J., et al., 2012. Intranasal drug delivery: anefficient and non-invasive route for systemic administration: focus on opioids. Pharmacol. Ther. 134, 366–379, 2012.

Hao, X.M., Li, L.D., Duan, C.L., Li, Y.J., 2017. Neuroprotective effect of alpha-mangostin on mitochondrial dysfunction and alpha-synuclein aggregation in rotenone-induced model of Parkinson's disease in differentiated SH-SY5Y cells. J. Asian Nat. Prod. Res. 19, 833–845.

Haque, S., Md, S., Sahni, J.K., Ali, J., Baboota, S., 2014. Development and evaluation of brain targeted intranasal alginate nanoparticles for treatment of depression. J. Psychiatr. Res. 48, 1–12.

Henriksen, I., Green, K.L., Smart, J.D., Smistad, G., Karlsen, J., 1996. Bioadhesion ofhydrated chitosans: an in vitro and in vivo study. Int. J. Pharm. 145, 231–240.

Heurtault, B., Frisch, B., Pons, F., 2010. Liposomes as delivery systems for nasal vaccination: strategies and outcomes. Expert Opin. Drug Deliv. 7, 829–844.

Herring, W.J., Assaid, C., Budd, K., Vargo, R., Mazenko, R.S., Lines, C., et al., 2017. A phase Ib randomized controlled study to evaluate the effectiveness of a single-dose of the NR2B selective N-Methyl-D-Aspartate antagonist MK-0657 on levodopa-induced dyskinesias and motor symptoms in patients with Parkinson disease. Clin. Neuropharmacol. 40, 255–260.

Horvat, S., Feher, A., Wolburg, H., Sipos, P., Veszelka, S., Toth, A., et al., 2009. Sodium hyaluronate as a mucoadhesive component in nasal formulation enhances delivery of molecules to brain tissue. Eur. J. Pharm. Biopharm. 72, 252–259.

Illum, L., 2002. Nasal drug delivery: new developments and strategies. Drug Discov. Today 7, 1184–1189.

Jafarieh, O., Md, S., Ali, M., Baboota, S., Sahni, J.K., Kumari, B., et al., 2015. Design, characterization, and evaluation of intranasal delivery of ropinirole-loaded mucoadhesive nanoparticles for brain targeting. Drug Dev. Ind. Pharm. 41, 1674–1681.

Ji, B., Maeda, J., Higuchi, M., Inoue, K., Akita, H., Harashima, H., et al., 2006. Pharmacokinetics and brain uptake of lactoferrin in rats. Life Sci 78, 851–855.

Karmakar, A., Zhang, Q., Zhang, Y., 2014. Neurotoxicity of nanoscale materials. J. Food Drug Anal. 22, 147–160.

Kanazawa, T., 2015. Brain delivery of small interfering ribonucleic acid and drugs through intranasal administration with nano-sized polymer micelles. Med Devices 8, 57–64.

Khan, A., Imam, S.S., Aqil, M., Ahad, A., Sultana, Y., Ali, A., et al., 2016. Brain targeting of temozolomide via the intranasal route using lipid-based nanoparticles: brain pharmacokinetic and scintigraphic analyses. Mol. Pharm. 13, 3773–3782.

Kreuter, J., 2001. Nanoparticulate systems for brain delivery of drugs. Adv. Drug Deliv. Rev. 47, 65–81.

Kurti, L., Gaspar, R., Marki, A., Kapolna, E., Bocsik, A., Veszelka, S., et al., 2013. In vitro and in vivo characterization of meloxicam nanoparticles designed for nasal administration. Eur. J. Pharm. Sci. 50, 86–92.

Kulkarni, A.D., Vanjari, Y.H., Sancheti, K.H., Belgamwar, V.S., Surana, S.J., Pardeshi, C.V., 2015. Nanotechnology-mediated nose to brain drug delivery for Parkinson's disease: a mini review. J. Drug Target. 23, 775–788.

Kumar, M., Pandey, R.S., Patra, K.C., Jain, S.K., Soni, M.L., Dangi, J.S., et al., 2013. Evaluation of neuropeptide loaded trimethyl chitosan nanoparticles for nose to brain delivery. Int. J. Biol. Macromol. 61, 189–195.

Leyva-Gomez, G., Cortes, H., Magana, J.J., Leyva-Garcia, N., Quintanar-Guerrero, D., Floran, B., 2015. Nanoparticle technology for treatment of Parkinson's disease: the role of surface phenomena in reaching the brain. Drug Discov. Today 20, 824–837.

Lalani, J., Baradia, D., Lalani, R., Misra, A., 2014. Brain targeted intranasal delivery of tramadol: comparative study of microemulsion and nanoemulsion. Pharm. Dev. Technol. 1–10.

Lemarchand, C., Gref, R., Couvreur, P., 2004. Polysaccharide-decorated nanoparticles. Eur. J. Pharm. Biopharm. 58, 327–341.

Li, X., Zhang, R., Liang, R., Liu, W., Wang, C., Su, Z., et al., 2014. Preparation and characterization of sustained-release rotigotine film-forming gel. Int. J. Pharm. 460, 273–279.

Lucchini, R.G., Dorman, D.C., Elder, A., Veronesi, B., 2012. Neurological impacts from inhalation of pollutants and the nose-brain connection. NeuroToxicol. 33, 838–841.

Marttin, E., Schipper, N.G.M., Coos Verhoef, J., Merkus, F.W.H.M., 1998. Nasal mucociliary clearance as a factor in nasal drug delivery. Adv. Drug Deliv. Rev. 29, 13–38.

Mittal, D., Md, S., Hasan, Q., Fazil, M., Ali, A., Baboota, S., et al., 2016. Brain targeted nanoparticulate drug delivery system of rasagiline via intranasal route. Drug Deliv. 23, 130–139.

Mo, X., Liu, E., Huang, Y., 2019. Chapter 16—the intra-brain distribution of brain targeting delivery systems. In: Gao, H., Gao, X. (Eds.), Brain Targeted Drug Delivery System. Academic Press, Cambridge, MA, USA, pp. 409–438.

Md, S., Haque, S., Sahni, J.K., Baboota, S., Ali, J., 2011. New non-oral drug delivery systems for Parkinson's disease treatment. Expert Opin. Drug Deliv. 8, 359–374.

Mistry, A., Stolnik, S., Illum, L., 2015. Nose-to-Brain delivery: investigation of the transport of nanoparticles with different surface characteristics and sizes in excised porcine olfactory epithelium. Mol. Pharm. 12, 2755–2766.

Md, S., Khan, R.A., Mustafa, G., Chuttani, K., Baboota, S., Sahni, J.K., et al., 2013. Bromocriptine loaded chitosan nanoparticles intended for direct nose to brain delivery: pharmacodynamic, Pharmacokinetic and Scintigraphy study in mice model. Eur. J. Pharm. Sci. 48, 393–405.

Mittal, D., Md, S., Hasan, Q., Fazil, M., Ali, A., Baboota, S., et al., 2014. Brain targeted nanoparticulate drug delivery system of rasagiline via intranasal route. Drug Deliv. 23, 130–139.

Muntimadugu, E., Dhommati, R., Jain, A., Challa, V.G., Shaheen, M., Khan, W., 2016. Intranasal delivery of nanoparticle encapsulated tarenflurbil: a potential brain targeting strategy for Alzheimer's disease. Eur. J. Pharm. Sci. 92, 224–234.

Nageeb El-Helaly, S., Abd Elbary, A., Kassem, M.A., El-Nabarawi, M.A., 2017. Electrosteric stealth Rivastigmine loaded liposomes for brain targeting: preparation, characterization, ex vivo, bio-distribution and in vivo pharmacokinetic studies. Drug Deliv. 24, 692–700.

Nagpal, K., Singh, S.K., Mishra, D.N., 2010. Chitosan nanoparticles: a promising system in novel drug delivery. Chem. Pharm. Bull. 58, 1423–1430.

Neuwelt, E.A., Bauer, B., Fahlke, C., Fricker, G., Iadecola, C., Janigro, D., et al., 2011. Engaging neuroscience to advance translational research in brain barrier biology. Nat. Rev. Neurosci. 12, 169–182.

Ogawa, T., Oyama, G., Hattori, N., 2018. Transdermal rotigotine patch in Parkinson's disease with a history of intestinal operation. BMJ Case Rep. 2018, pii: bcr-2017-223722.

Ozsoy, Y., Gungor, S., Cevher, E., 2009. Nasal delivery of high molecular weight drugs. Molecules 14, 3754–3779.

Pozzoli, M., Rogueda, P., Zhu, B., Smith, T., Young, P.M., Traini, D., et al., 2016. Dry powder nasal drug delivery: challenges, opportunities and a study of the commercial Teijin Puvlizer Rhinocort device and formulation. Drug Dev. Ind. Pharm. 42, 1660–1668.

Pardeshi, C.V., Belgamwar, V.S., 2013. Direct nose to brain drug delivery via integrated nerve pathways bypassing the blood-brain barrier: an excellent platform for brain targeting. Expert Opin. Drug Deliv. 10, 957–972.

Qian, Z.M., Wang, Q., 1998. Expression of iron transport proteins and excessive iron accumulation in the brain in neurodegenerative disorders. Brain Res Brain Res Rev 27, 257–267.

Qiang, F., Shin, H.J., Lee, B.J., Han, H.K., 2012. Enhanced systemic exposure of fexofenadine via the intranasal administration of chitosan-coated liposome. Int. J. Pharm. 430, 161–166.

Quintana, D.S., Guastella, A.J., Westlye, L.T., Andreassen, O.A., 2016. The promise and pitfalls of intranasally administering psychopharmacological agents for the treatment of psychiatric disorders. Mol. Psychiatry 29–38.

Raj, R., Wairkar, S., Sridhar, V., Gaud, R., 2017. Pramipexole dihydrochloride loaded chitosan nanoparticles for nose to brain delivery: development, characterization and in vivo anti-Parkinson activity. Int. J. Biol. Macromol. 109, 27–35.

Sonvico, F., Clementino, A., Buttini, F., Colombo, G., Pescina, S., Staniscuaski Guterres, S., et al., 2018. Surface-modified nanocarriers for nose-to-brain delivery: from bioadhesion to targeting. Pharmaceutics 10, E34 pii.

Sukowski, W.W., Pentak, D., Nowak, K., Sukowska, A., 2005. The influence of temperature, cholesterol content and pH on liposome stability. J. Mol. Struct. 744–747, 737–747.

Sridhar, V., Gaud, R., Bajaj, A., Wairkar, S., 2018. Pharmacokinetics and pharmacodynamics of intranasally administered selegiline nanoparticleswith improved brain delivery in Parkinson's disease. Nanomedicine 14, 2609–2618.

Soppimath, K.S., Aminabhavi, T.M., Kulkarni, A.R., Rudzinski, W.E., 2001. Biodegradable polymeric nanoparticles as drug delivery devices. J. Control. Release 70, 1–20.

Simko, M., Mattsson, M.O., 2010. Risks from accidental exposures to engineered nanoparticles and neurological health effects: a critical review. Part. Fibre Toxicol 7, 42.

Tafaghodi, M., Sajadi Tabassi, S.A., Jaafari, M.R., 2006. Induction of systemic and mucosal immune responses by intranasal administration of alginate microspheres encapsulated with tetanus toxoid and CpG-ODN. Int. J. Pharm. 319, 37–43.

Teleanu, D.M., Chircov, C., Grumezescu, A.M., Volceanov, A., Teleanu, R.I., 2018. Blood-brain delivery methods using nanotechnology. Pharmaceutics 10, E269 pii.

Thorne, R.G., Pronk, G.J., Padmanabhan, V., Frey, W.H., 2004. Delivery of insulin-like growth factor-I to the rat brain and spinal cord along olfactory and trigeminal pathways following intranasal administration. Neuroscience 127, 481–496.

Tzeyung, A.S., Md, S., Bhattamisra, S.K., Madheswaran, T., Alhakamy, N.A., Aldawsari, H.M., et al., 2019. Fabrication, optimization, and evaluation of rotigotine-loaded chitosan nanoparticles for nose-to-brain delivery. Pharmaceutics 11, E26 pii.

Ugwoke, M.I., Agu, R.U., Verbeke, N., Kinget, R., 2005. Nasal mucoadhesive drug delivery: background, applications, trends and future perspectives. Adv. Drug Deliv. Rev. 57, 1640–1665.

Vaka, S.R., Shivakumar, H.N., Repka, M.A., Murthy, S.N., 2013. Formulation and evaluation of carnosic acid nanoparticulate system for upregulation of neurotrophins in the brain upon intranasal administration. J. Drug Target. 21, 44–53.

Verma, A., Stellacci, F., 2010. Effect of surface properties on nanoparticle-cell interactions. Small 6, 12–21.

Wang, S., Chen, P., Zhang, L., Yang, C., Zhai, G., 2012. Formulation and evaluation of microemulsion-based in situ ion-sensitive gelling systems for intranasal administration of curcumin. J. Drug Target. 20, 831–840.

Ward, P.P., Uribe-Luna, S., Conneely, O.M., 2002. Lactoferrin and host defense. Biochem. Cell Biol. 80, 95–102.

Win-Shwe, T.T., Sone, H., Kurokawa, Y., Zeng, Y., Zeng, Q., Nitta, H., et al., 2014. Effects of PAMAM dendrimers in the mouse brain after a single intranasal instillation. Toxicol. Lett. 228, 207–215.

Wang, Z.Y., Sreenivasmurthy, S.G., Song, J.X., Liu, J.Y., Li, M., 2018. Strategies for brain-targeting liposomal delivery of small hydrophobic molecules in the treatment of neurodegenerative diseases. Drug Discov. Today 24, 595–605.

Wong, L.R., Ho, P.C., 2018. Role of serum albumin as a nanoparticulate carrier for nose-to-brain delivery of R-flurbiprofen: implications for the treatment of Alzheimer's disease. J. Pharm. Pharmacol. 70, 59–69.

Yan, X., Xu, L., Bi, C., Duan, D., Chu, L., Yu, X., et al., 2018. Lactoferrin-modified rotigotine nanoparticles for enhanced nose-to-brain delivery: LESA-MS/MS-based drug biodistribution, pharmacodynamics, and neuroprotective effects. Int. J. Nanomed. 13, 273–281.

Further reading

Raj, R., Wairkar, S., Sridhar, V., Gaud, R., 2018. Pramipexole dihydrochloride loaded chitosan nanoparticles for nose to brain delivery: development, characterization and in vivo anti-Parkinson activity. Int. J. Biol. Macromol. 109, 27–35.

CHAPTER 9

α-Synuclein-targeted nanoparticles

1. Introduction

1.1 α-Synuclein

α-Synuclein is found in both Alzheimer's disease (AD) and Parkinson's disease (PD). However, in PD, it is found in more clump-like structures, thus indicating the destabilization of clumps as a primary therapeutic target (Wakabayashi et al., 2007; Broderick et al., 2017). α-Synuclein is a cryptic protein with 142 amino acids and profoundly found in dentate gyrus, hippocampus, olfactory lobe, amygdale, cerebellum, thalamus, and presynaptic terminals (Lim et al., 2011; Khan et al., 2018). It comprises a central region, an N-terminal repeat region, and a C-terminal region (Khan et al., 2018). N-terminal is a positively charged region containing repetitive seven residues and fatty acid—binding protein (FABP). FABP attaches to lipophilic micelles, in turn forming helical structure (Sode et al., 2006; Khan et al., 2018). C-terminal is negatively charged and without folding in structure and avoids accumulation of α-synuclein and its communication with N-terminal or nonamyloid β component (NAC) region and restores the unfolded structure of α-synuclein (Emamzadeh, 2016; Khan et al., 2018). α-Synuclein plays an essential role in conservation of neural plasticity, dopamine synthesis, discharge of pulse transmitting synaptic chemicals, and development and functions of neurons (Khan et al., 2018). It is delivered by glial cells into the extracellular matrix and absorbed by surrounding neuronal cells, in turn contributing for the pathogenesis of the disease (Khan et al., 2018). It has been found that α-synuclein fibril attached to heparan sulfate proteoglycans component of cell membrane facilitates its entry into the cell by endocytosis (Holmes et al., 2013; Khan et al., 2018). A few factors that influence α-synuclein aggregation include forceful shaking, existence of beads (Gray et al., 2011) and/or α-synuclein seeds (Luk et al., 2012), environmental changes (cohlberg et al., 2002, Yamin

Parkinson's Disease Therapeutics
ISBN 978-0-12-819882-7
https://doi.org/10.1016/B978-0-12-819882-7.00009-X

et al., 2003; Munishkina et al., 2003), and acidic conditions and enhanced temperature (Sardar Sinha et al., 2018). Accumulation of misfolded α-synuclein clumps in the neurons, microglia, or nerve fibers drive toward the pathogenesis of PD (Burre et al., 2018; Khan et al., 2018). It also contributes for brain disorders, which are called synucleinopathies (Schommer et al., 2018). Although extensive research has been done, mechanism of α-synuclein aggregation is elusive (Schommer et al., 2018). α-Synuclein is broken down by proteasome (Webb et al., 2003, Bennett et al., 1999, Alvarez-Castelao et al., 2014) and proteasomal impairment has been observed in PD (Mc Naught et al., 2001, 2006; Schommer et al., 2018) .

2. 1-Methyl-4-phenyl-1,2,3,6-tetrahydropyridine treatment

1-Methyl-4-phenyl-1,2,3,6-tetrahydropyridine (MPTP) treatment initiates the translocation of α-synuclein from usual synaptic region to deteriorating cell bodies as aggregates (Kowall et al., 2000; Dauer and Przedborski, 2003; Zhang et al., 2018). α-Synuclein aggregation induced by MPTP demonstrates the initial stages of Lewy body formation and plays an essential role in pathogenesis of PD (Fornai et al., 2005; Zhang et al., 2018). In line with this, chronic treatment with MPTP initiates aggregation and nitration of α-synuclein in the cytosol of substantia nigra pars compacta (SNpc) dopaminergic neurons (Vila et al., 2000; Meredith et al., 2008; Gibrat et al., 2009; Zhang et al., 2018). Therefore, MPTP models have been extensively used in pathology and therapeutic studies of PD.

Snca gene encodes α-synuclein and it was found that mutation in this gene results in changed expression of protein and enhanced the PD incidence (Khan et al., 2018). Contrary to this, a few studies showed ambiguity regarding the role of α-synuclein in PD pathogenesis (Brundin et al., 2017; Khan et al., 2018). In line with this, stress, antioxidant defense, protein metabolism, and enhanced energy demand also initiate α-synuclein accumulation but do not contribute for PD (Brundin et al., 2017; Khan et al., 2018). Wealth of studies has shown that α-synuclein accumulates at synaptic terminals and attaches to synaptic vesicles and other cell membranes, thus influencing functional activity and accumulation tendency (Sardar Sinha et al., 2018). Posttranslational modifications of α-synuclein induced by oxidative stress such as alteration by 4-hydroxy-2-nonenal (HNE), nitration, and oxidation enhance oligomerization process (Schmid et al., 2013; Xiang et al., 2013; Sardar Sinha et al., 2018). Current treatment aimed at α-synuclein includes decreasing production, accumulation, and enhancement of both intra- and extracellular degradation of α-synuclein (Khan et al., 2018).

3. Fibrillation

Monomeric α-synuclein aggregates into toxic fibrillary species, which induce neuronal dysfunction by deteriorating mitochondrial function and degradation mechanisms (Mazzulli et al., 2016; Fusco et al., 2017; Ludtmann et al., 2018; Grassi et al., 2018; Mohammad-Beigi et al., 2019). As fibrillation is an intricate process that involves multifarious biological aspects and conditions resulting in various types of self-assembly states, the approach to curtail α-synuclein—induced toxicity has been a herculean task until now (Semerdzhiev et al., 2014; Buell et al., 2014; Galvagnion et al., 2015; Flagmeier et al., 2016; Iljina et al., 2018; Mohammad-Beigi et al., 2019). The intense hydrophobicity of NAC region promotes fibrillation process (Geodert, 2001; Mohammad-Beigi et al., 2019). Therefore, there has been a growing need to develop therapeutics to eliminate fibrils or stop fibril formation.

4. Early diagnosis

Similarity between symptoms of PD and other diseases such as essential tremor and Parkinsonism syndrome makes the PD diagnosis a difficult task (Gao et al., 2019). In addition, treating PD after the onset is even more difficult because attenuating the disease progress is a major biological riddle (Lees et al., 2009; Gao et al., 2019). As cerebrospinal fluid—based α-synuclein analysis is an invasive procedure, wealth of studies was targeted at blood-based α-synuclein for PD diagnosis. However, they showed controversial results and limited success (Duran et al., 2010; Mata et al., 2010; Gao et al., 2019). To overcome the abovementioned challenges and to develop a robust PD diagnostic tool, specific peptoids were used. A recently designed peptoid, α-synuclein binding peptoid (ASBP-7), with its substantial affinity and specificity to α-synuclein appropriately identifies PD sera from the normal by detecting α-synuclein in it (Gao et al., 2019).

4.1 Therapy

Studies have shown that introduction of siRNA in squirrel monkeys decreased α-synuclein level by 40%—50% compared to control group

(McCormack et al., 2001, Khan et al., 2018). Similarly, in another study, rats treated with shRNA showed 35% decrease in α-synuclein expression (Zharikov et al., 2015; Khan et al., 2018). Regular physiological activity of α-synuclein could be reestablished by attenuating its accumulation and misfolding (Khan et al., 2018). Heat shock proteins (HSPs) successfully eliminate protein aggregates, which play role in neurological disorders such as AD and PD (Gorenberg and Chandra, 2017; Khan et al., 2018). Autophagy attenuators such as trehalose enhanced the elimination of protein aggregates via lysosomal biogenesis but neuronal toxicity impeded its success (Ghavami et al., 2014; Khan et al., 2018).

27-Hydroxycholesterol and heat shock protein 70 (HSP70) being the chief regulators of α-synuclein can be opted to develop novel therapeutics (Schommer et al., 2018). α-Synuclein attaches itself to HSP70 and undergoes lysis by the ubiquitin proteasomal system and autophagy (Webb et al., 2003; Dedmon et al., 2005; Luk et al., 2008; Aprile et al., 2017). Several lines of evidence have proven that hyperlipidemia enhances the risk of PD (Mutez et al., 2009; Gao et al., 2012; Marwarha and Ghribi, 2015; Schommer et al., 2018). Conversely, a few studies have shown reduced risk (Simon et al., 2007; Powers et al., 2009; Miyake et al., 2010) or no relation between increased cholesterol levels and PD (Abbott et al., 2003; de Lau et al., 2005; Schommer et al., 2018).

4.1.1 Nanotechnology

4.1.1.1 Nanotechnology for Parkinson's disease diagnosis

To have early and appropriate PD diagnosis, Au-doped titanium dioxide (TiO_2) nanotube arrays were designed with enhanced sensitivity photoelectrochemical immunosensor to identify α-synuclein (An et al., 2010; Gunay et al., 2016). In another study, nanomanipulation of single molecule of α-synuclein was used to detect misfolding and self-assembly of α-synuclein (Yu and Lyubchenko, 2009; Gunay et al., 2016). In another study, quantitative analysis of PD-related neurotransmitters was accomplished using plasmon absorbance and gold nanoparticles (AuNPs), and accurate results were obtained (Baron et al., 2005; Gunay et al., 2016) (see Table 9.1).

4.1.2 Nanotechnology for Parkinson's disease therapy

AuNPs have been used as nonviral gene vector to transport siRNA, which profoundly downregulates α-synuclein expressing genes (Khan et al., 2018). Nanoformulation localized plasmid DNA (pDNA) into cells via receptor-mediated endocytosis and attenuated apoptosis in PC12 cells

Table 9.1 Nanoparticles (NPs) targeted at Parkinson's disease (PD) diagnosis and therapy.

Nanoformulation	PD diagnosis	Reference
Au-doped TiO_2 nanotube arrays	Identification of α-synuclein	An et al. (2010)
AuNPs	PD-related neurotransmitter analysis	Baron et al. (2005)
Nanoformulation	**Therapeutic Effect**	**Reference**
AuNPs	Transport of siRNA and downregulation of α-synuclein genes	Khan et al. (2018)
Lipid vesicles	Prolonged the accumulation of 4-hydroxy-2-nonenal−conjugated α-synuclein	Sardar Sinha et al. (2018)
Polybutylcyanoacrylate (PBCA) NPs	α-Synuclein elimination	Hasadsri et al. (2009) Klyachko et al. (2014) Paurush and Daniel (2015) Siddiqui et al. (2018)
mAb-tethered PBCA NPs	Neuroprotection in cultured neurons	Hasadsri et al. (2009), Gunay et al. (2016)
Positively charged poly(allylamine hydrochloride) conjugated AuNPs	Absorption of α-synuclein	Yang et al. (2014)
Polyamidoamine dendrimers	Breakdown of preexisting α-synuclein fibrils and attenuated the β-sheet formation	Rekas et al. (2009), Gunay et al. (2016)
Viologen-phosphorus dendrimers	Attenuated α-synuclein formation	Milowska et al. (2013)
Curcumin-loaded polysorbate-80−conjugated cerasome	Ameliorated motor behavior, dopamine levels, tyrosine hydroxylase expression, and α-synuclein scavenging	Zhang et al. (2018)
Nanoobjects with positively and negatively charged graphene sheets or superparamagnetic iron oxide NPs	Inhibited fibrillation	Mohammad-Beigi et al. (2019)

(Hu et al., 2018; Khan et al., 2018). Lipid vesicles considerably prolonged the accumulation of HNE-conjugated α-synuclein (Sardar Sinha et al., 2018). Anti—α-synuclein tethered polybutylcyanoacrylate (PBCA) NPs facilitate α-synuclein elimination [Hasadsri et al., 2009a,b; Klyachko et al., 2014; Paurush and Daniel, 2015; Siddiqui et al., 2018). α-Synuclein—specific monoclonal antibody (mAb)—tethered PBCA NPs showed enhanced internalization by endocytosis and neuroprotection in cultured neurons (Hasadsri et al., 2009a,b; Gunay et al., 2016). Studies have also shown the substantial ability of positively charged poly(allylamine hydrochloride) conjugated AuNPs in the absorption of α-synuclein (Yang et al., 2014; Gunay et al., 2016).

4.1.3 Dendrimers

Rekas et al. (2009) designed polyamidoamine dendrimers (generations G3, Gr, and G5) to overcome α-synuclein fibrillation. These dendrimers substantially initiated the breakdown of preexisting α-synuclein fibrils and attenuated the β-sheet formation (Gunay et al., 2016). In another study, viologen-phosphorus dendrimers were developed to overcome α-synuclein fibrillation (Gunay et al., 2016). These dendrimers with phosphonate groups on the surface successfully attenuated α-synuclein formation (Milowska et al., 2013; Gunay et al., 2016).

4.1.4 Curcumin and curcumin-loaded nanoparticles

Curcumin renders neuroprotection with regular and adequate intake (Gota et al., 2010; Mythri et al., 2012; Fu et al., 2015; Zhang et al., 2018). Mounting evidence has shown that it substantially attenuates and scavenges α-synuclein (Shrikanth Gadad et al., 2012; Jiang et al., 2013; Zhang et al., 2018). Nevertheless, low bioavailability due to low absorption increased metabolism and removal from systemic circulation and low BBB spanning ability impede its success (Anand et al., 2007; Nelson et al., 2017; Zhang et al., 2018). Therefore, to circumvent these biological riddles, curcumin-loaded polysorbate-80 (PS80)—conjugated cerasome (CPC) was designed. CPC ameliorated motor behavior, dopamine levels, tyrosine hydroxylase expression, and α-synuclein scavenging in MPTP-treated PD mice (Zhang et al., 2018). PS80 functionalization significantly increased BBB spanning of CPC by transcytosis (Zhang et al., 2018).

4.1.5 Nanoparticles targeted at α-synuclein fibrillation

α-Synuclein aggregation is an electrostatically initiated process, which involves dipole—dipole interactions and further stabilizations van der waals interactions and hydrogen bonds (Mohammad-Beigi et al., 2019). To overcome this aggregation process, nanoobjects were designed in which positively and negatively charged graphene sheets or superparamagnetic iron oxide NPs connect with α-synuclein's N-terminal or C-terminal charged moieties and with hydrophobic moieties in the nonamyloid component (61−95) region to inhibit fibrillation (nucleation and elongation) (Mohammad-Beigi et al., 2019). Although several earlier studies showed initiation of α-synuclein fibrillation (gold and surface-functionalized albumin NPs) (Alvarez et al., 2013; Mohammad-Beigi et al., 2015), yet a few studies showed attenuation of α-synuclein fibrillation and decrease preformed fibrils, thus protecting neurons (Kim et al., 2018; Mohammad-Beigi et al., 2019).

5. Conclusions and future perspectives

The potential common factor between both AD and PD is Lewy body protein accumulation. As several similarities between AD and PD were found, PD condition probably results in the onset of AD also (Dugger et al., 2012, Broderick et al., 2017). Therefore, there is a possibility that the therapeutic strategies designed to target α-synuclein accumulation also show potential to combat other diseases also. Au-doped titanium dioxide (TiO_2) nanotube arrays play an essential role in PD diagnosis. Several natural compounds and associated NPs also showed enhanced efficacy in attenuating α-synuclein accumulation and fibrillation. Although recent progress in PD theranostic avenues is promising, still there is a growing need to develop robust strategies.

References

Abbott, R.D., Webster Ross, G., White, L.R., Sanderson, W.T., Burchfiel, C.M., Kshon, M., et al., 2003. Environmental, life-style, and physical precursors of clinical Parkinson's disease: recent findings from the Honolulu-Asia Aging Study. J. Neurol. 250, 11130−11139.

Alvarez, Y.D., Fauerbach, J.A., Pellegrotti, J.S.V., Jovin, T.M., Jares-Erijman, E.A., Stefani, F.D., 2013. Influence of gold nanoparticles on the kinetics of α-synuclein aggregation. Nano Lett. 13, 6156−6163.

Alvarez-Castelao, B., Goethals, M., Vandekerckhove, J., Castano, J.G., 2014. Mechanism of cleavage of alpha-synuclein by the 20S proteasome and modulation of its degradation by the RedOx state of the N-terminal methionines. Biochim. Biophys. Acta 1843, 352−365.

An, Y., Tang, L., Jiang, X., Chen, H., Yang, M., Jin, L., et al., 2010. A photoelectrochemical immunosensor based on Au-doped TiO2 nanotube arrays for the detection of α-synuclein. Chemistry 16, 14439−14446.

Anand, P., Kunnumakkara, A.B., Newman, R.A., Aggarwal, B.B., 2007. Bioavailability of curcumin: problems and promises. Mol. Pharm. 4, 807–818.

Aprile, F.A., Arosio, P., Fusco, G., Chen, S.W., Kumita, J.R., Dhulesia, A., et al., 2017. Inhibition of α-synuclein fibril elongation by Hsp70 is governed by a kinetic binding competition between α-synuclein species. Biochemistry 56, 1177–1180.

Baron, R., Zayats, M., Willner, I., 2005. Dopamine-, L-DOPA-, adrenaline-, and noradrenaline-induced growth of Au nanoparticles: assays for the detection of neurotransmitters and of tyrosinase activity. Anal. Chem. 77, 1566–1571.

Bennett, M.C., Bishop, J.F., Leng, Y., Chock, P.B., Chase, T.N., Mouradian, M.M., 1999. Degradation of alpha-synuclein by proteasome. J. Biol. Chem. 274, 33855–33858.

Broderick, P.A., Wenning, L., Li, Y.S., 2017. Neuromolecular imaging, a nanobiotechnology for Parkinson's disease: advancing pharmacotherapy for personalized medicine. J. Neural Transm. 124, 57–78.

Brundin, P., Dave, K.D., Kordower, J.H., 2017. Therapeutic approaches to target alpha-synuclein pathology. Exp. Neurol. 298, 225–235.

Buell, A.K., Galvagnion, C., Gaspar, R., Sparr, E., Vendruscolo, M., Knowles, T.P., et al., 2014. Solution conditions determine the relative importance of nucleation and growth processes in α-synuclein aggregation. Proc. Natl. Acad. Sci. U.S.A. 111, 7671–7676.

Burre, J., Sharma, M., Sudhof, T.C., 2018. Cell biology and pathophysiology of alpha-Synuclein. Cold Spring Harb Perspect Med 8, a024091 pii.

Cohlberg, J.A., Li, J., Uversky, V.N., Fink, A.L., 2002. Heparin and other glycosaminoglycans stimulate the formation of amyloid fibrils from alpha-synuclein in vitro Biochemistry 41, pp. 1502–1511.

Dauer, W., Przedborski, S., 2003. Parkinson's disease: mechanisms and models. Neuron 39, 889–909.

de Lau, L.M.L., Bornebroek, M., Witteman, J.C.M., Hofman, A., Koudstaal, P.J., Breteler, M.M.B., 2005. Dietary fatty acids and the risk of Parkinson disease: the Rotterdam study. Neurology 64, 2040–2045.

Dedmon, M.M., Christodoulou, J., Wilson, M.R., Dobson, C.M., 2005. Heat shock protein70 inhibits alpha-synuclein fibril formation via preferential binding to prefibrillar species. J. Biol. Chem. 280, 14733–14740.

Dugger, B.N., Serrano, G.E., Sue, L.I., Walker, D.G., Adler, C.H., Shill, H.A., 2012. Presence of striatal amyloid plaques in Parkinson's disease dementia predicts concomitant Alzheimer's disease: usefulness for amyloid imaging. J. Parkinson's Dis. 2, 57–65.

Duran, R., Barrero, F.J., Morales, B., Luna, J.D., Ramirez, M., Vives, F., 2010. Plasma alpha-synuclein in patients with Parkinson's disease with and without treatment. Mov. Disord. 25, 489–493.

Emamzadeh, F.N., 2016. Alpha-synuclein structure, functions, and interactions. J. Res. Med. Sci. 21, 1735–1995.

Flagmeier, P., Meisl, G., Vendruscolo, M., Knowles, T.P., Dobson, C.M., Buell, A.K., et al., 2016. Mutations associated with familial Parkinson's disease alter the initiation and amplification steps of α-synuclein aggregation. Proc. Natl. Acad. Sci. U. S. A. 113, 10328–10333.

Fornai, F., Schluter, O.M., Lenzi, P., Gesi, M., Ruffoli, R., Ferrucci, M., 2005. Parkinson-like syndrome induced by continuous MPTP infusion: convergent roles of the ubiquitin-proteasome system and α-synuclein. Proc. Natl. Acad. Sci. U. S. A. 102, 3413–3418.

Fu, W., Zhuang, W., Zhou, S., Wang, X., 2015. Plant-derived neuroprotective agents in Parkinson's disease. Am J Transl Res 7, 1189–1202.

Fusco, G., Chen, S.W., Williamson, P.T., Cascella, R., Perni, M., Jarvis, J.A., et al., 2017. Structural basis of membrane disruption and cellular toxicity by α-synuclein oligomers. Science 358, 1440–1443.

Gadad, B.S., Subramanya, P.K., Pullabhatla, S., Shantharam, I.S., Rao, K.S., 2012. Curcumin-glucoside, a novel synthetic derivative of curcumin, inhibits α-synuclein oligomer formation: relevance to Parkinson's disease. Curr. Pharmaceut. Des. 18, 76–84.

Galvagnion, C., Buell, A.K., Meisl, G., Michaels, T.C., Vendruscolo, M., Knowles, T.P., et al., 2015. Lipid vesicles trigger α-synuclein aggregation by stimulating primary nucleation. Nat. Chem. Biol. 11, 229–234.

Gao, X., Simon, K.C., Schwarzschild, M.A., Ascherio, A., 2012. Prospective study of statin use and risk of Parkinson disease. Arch. Neurol. 69, 380–384.

Gao, H., Zhao, Z., He, Z., Wang, H., Liu, M., Hu, Z., et al., 2019. Detection of Parkinson's disease through the peptoid recognizing α-synuclein in serum. ACS Chem. Neurosci. 10, 1204–1208.

Ghavami, S., Shojaei, S., Yeganeh, B., Ande, S.R., Jangamreddy, J.R., Mehrpour, M., et al., 2014. Autophagy and apoptosis dysfunction in neurodegenerative Disorders. Prog. Neurobiol. 112, 24–49.

Gibrat, C., Saint-Pierre, M., Bousquet, M., Levesque, D., Rouillard, C., Cicchetti, F., 2009. Differences between subacute and chronic MPTP mice models: investigation of dopaminergic neuronal degeneration and α-synuclein inclusions. J. Neurochem. 109, 1469–1482.

Goedert, M., 2001. Alpha-Synuclein and neurodegenerative diseases. Nat. Rev. Neurosci. 2, 492–501.

Gorenberg, E.L., Chandra, S.S., 2017. The role of co-chaperones in synaptic proteostasis and neurodegenerative disease. Front. Neurosci. 11, 248.

Gota, V.S., Maru, G.B., Soni, T.G., Gandhi, T.R., Kochar, N., Agarwal, M.G., 2010. Safety and pharmacokinetics of a solid lipid curcumin particle formulation in osteosarcoma patients and healthy volunteers. J. Agric. Food Chem. 58, 2095–2099.

Grassi, D., Howard, S., Zhou, M., Diaz-Perez, N., Urban, N.T., Guerrero-Given, D., et al., 2018. Identification of a highly neurotoxic α-synuclein species inducing mitochondrial damage and mitophagy in Parkinson's disease. Proc. Natl. Acad. Sci. U. S. A. 115, E2634–E2643.

Grey, M., Linse, S., Nilsson, H., Brundin, P., Sparr, E., 2011. Membrane interaction of alphasynuclein in different aggregation states. J. Parkinson's Dis. 1, 359–371.

Gunay, M.S., Ozer, A.Y., Chalon, S., 2016. Drug delivery systems for imaging and therapy of Parkinson's disease. Curr. Neuropharmacol. 14, 376–391.

Hasadsri, L., Kreuter, J., Hattori, H., Iwasaki, T., George, J.M., 2009a. Functional protein delivery into neurons using polymeric nanoparticles. J. Biol. Chem. 284, 6972–6981.

Hasadsri, L., Kreuter, J., Hattori, H., Iwasaki, T., George, J.M., 2009b. Functional protein delivery into neurons using polymeric nanoparticles. J. Biol. Chem. 284, 6972–6981.

Holmes, B.B., DeVos, S.L., Kfoury, N., Li, M., Jacks, R., Yanamandra, K., et al., 2013. Diamond, Heparan sulfate proteoglycans mediate internalization and propagation of specific proteopathic seeds. Proc. Natl. Acad. Sci. U. S. A. 110, E3138–E3147.

Hu, K., Chen, X., Chen, W., Zhang, L., Li, J., Ye, J., et al., 2018. Guan, Neuroprotective effect of gold nanoparticles composites in Parkinson's disease model. Nanomedicine 14, 1123–1136.

Iljina, M., Dear, A.J., Garcia, G.A., De, S., Tosatto, L., Flagmeier, P., et al., 2018. Quantifying Co-oligomer formation by α-synuclein. ACS Nano 12, 10855–10866.

Jiang, T.F., Zhang, Y.J., Zhou, H.Y., Wang, H.M., Tian, L.P., Liu, J., et al., 2013. Curcumin ameliorates the neurodegenerative pathology in A53T α-synuclein cell model of Parkinson's disease through the downregulation of mTOR/p70S6K signaling and the recovery of macroautophagy. J. Neuroimmune Pharmacol. 8, 356–369.

Khan, A.R., Yang, X., Fu, M., Zhai, G., 2018. Recent progress of drug nanoformulations targeting to brain. J. Control. Release 291, 37–64.

Kim, D., Yoo, J.M., Hwang, H., Lee, J., Lee, S.H., Yun, S.P., et al., 2018. Graphene quantum dots prevent α-synucleinopathy in Parkinson's disease. Nat. Nanotechnol. 13, 812–818.

Klyachko, N.L., Haney, M.J., Zhao, Y., Manickam, D.S., Mahajan, V., Suresh, P., et al., 2014. Macrophages offer a paradigm switch for CNS delivery of therapeutic proteins. Nanomedicine 9, 403–422.

Kowall, N.W., Hantraye, P., Brouillet, E., Beal, M.F., McKee, A.C., Ferrante, R.J., 2000. MPTP induces alpha-synuclein aggregation in the substantia nigra of baboons. Neuroreport 11, 211–213.

Lees, A.J., Hardy, J., Revesz, T., 2009. Parkinson's disease. Lancet 373, 2055–2066.

Lim, Y., Kehm, V.M., Lee, E.B., Soper, J.H., Li, C., Trojanowski, J.Q., et al., 2011. alpha-Syn suppression reverses synaptic and memory defects in a mouse model of dementia with Lewy bodies. J. Neurosci. 31, 10076–10087.

Ludtmann, M.H., Angelova, P.R., Horrocks, M.H., Choi, M.L., Rodrigues, M., Baev, A.Y., et al., 2018. α-Synuclein oligomers interact with ATP synthase and open the permeability transition pore in Parkinson's disease. Nat. Commun. 9, 2293.

Luk, K.C., Mills, I.P., Trojanowski, J.Q., Lee, V.M.Y., 2008. Interactions between Hsp70 and the hydrophobic core of α-synuclein inhibit fibril assembly. Biochemistry 47, 12614–12625.

Luk, K.C., Kehm, V.M., Zhang, B., O'Brien, P., Trojanowski, J.Q., Lee, V.M., 2012. Intracerebral inoculation of pathological alpha-synuclein initiates a rapidly progressive neurodegenerative alpha-synucleinopathy in mice. J. Exp. Med. 209, 975–986.

Marwarha, G., Ghribi, O., 2015. Does the oxysterol 27-hydroxycholesterol underlie Alzheimer's disease-Parkinson's disease overlap? Exp. Gerontol. 68, 13–18.

Mata, I.F., Shi, M., Agarwal, P., Chung, K.A., Edwards, K.L., Factor, S.A., et al., 2010. SNCA variant associated with Parkinson disease and plasma alpha-synuclein level. Arch. Neurol. 67, 1350–1356.

Mazzulli, J.R., Zunke, F., Isacson, O., Studer, L., Krainc, D., 2016. α-Synuclein–Induced lysosomal dysfunction occurs through disruptions in protein trafficking in human midbrain synucleinopathy models. Proc. Natl. Acad. Sci. U. S. A. 113, 1931–1936.

McCormack, A.L., Mak, S.K., Henderson, J.M., Bumcrot, D., Farrer, M.J., Di McNaught, K.S.P., Olanow, C.W., Halliwell, B., Isacson, O., Jenner, P., 2001. Failure of the ubiquitin–proteasome system in Parkinson's disease. Nat. Rev. Neurosci. 2, 589–594.

McNaught, K.S.P., Jackson, T., JnoBaptiste, R., Kapustin, A., Olanow, C.W., 2006. Proteasomal dysfunction in sporadic Parkinson's disease. Neurology 66, S37–S49.

Meredith, G.E., Totterdell, S., Potashkin, J.A., Surmeier, D.J., 2008. Modeling PD pathogenesis in mice: advantages of a chronic MPTP protocol. Park. Relat. Disord. 14, S112–S115.

Milowska, K., Grochowina, J., Katir, N., El Kadib, A., Majoral, J.P., Bryszewska, M., et al., 2013. Viologen-phosphorus dendrimers inhibit α-synuclein fibrillation. Mol. Pharm. 10, 1131–1137.

Miyake, Y., Tanaka, K., Fukushima, W., Sasaki, S., Kiyohara, C., Tsuboi, Y., et al., 2010. Case-control study of risk of Parkinson's disease in relation to hypertension, hypercholesterolemia, and diabetes in Japan. J. Neurol. Sci. 293, 82–86.

Mohammad-Beigi, H., Shojaosadati, S.A., Marvian, A.T., Pedersen, J.N., Klausen, L.H., Christiansen, G., et al., 2015. Strong interactions with polyethylenimine-coated human serum albumin nanoparticles (PEIHSA NPs) alter α-synuclein conformation and aggregation kinetics. Nanoscale 7, 19627–19640.

Mohammad-Beigi, H., Hosseini, A., Adeli, M., Ejtehadi, M.R., Christiansen, G., Sahin, C., et al., 2019. Mechanistic understanding of the interactions between nano-objects with different surface properties and α-synuclein. ACS Nano 13, 3243–3256.

Munishkina, L.A., Phelan, C., Uversky, V.N., Fink, A.L., 2003. Conformational behavior and aggregation of alpha-synuclein in organic solvents: modeling the effects of membranes. Biochemistry 42, 2720–2730.

Mutez, E., Duhamel, A., Defebvre, L., Bordet, R., Destee, A., Kreisler, A., 2009. Lipid lowering drugs are associated with delayed onset and slower course of Parkinson's disease. Pharmacol. Res. 60, 41–45.

Mythri, R.B., Bharath, M.M., 2012. Curcumin: a potential neuroprotective agent in Parkinson's disease. Curr. Pharmaceut. Des. 18, 91–99.

Nelson, K.M., Dahlin, J.L., Bisson, J., Graham, J., Pauli, G.F., Walters, M.A., et al., 2017. The essential medicinal chemistry of curcumin. J. Med. Chem. 60, 1620–1637.

Paurush, A., Daniel, G.A., 2015. Nanotechnology in neurology: genesis, current status, and future prospects. Ann. Indian Acad. Neurol. 18, 382–386.

Powers, K.M., Smith-Weller, T., Franklin, G.M., Longstreth, W.T., Swanson, P.D., Checkoway, H., 2009. Dietary fats, cholesterol and iron as risk factors for Parkinson's disease. Park. Relat. Disord. 15, 47–52.

Rekas, A., Lo, V., Gadd, G.E., Cappai, R., Yun, S.I., 2009. PAMAM dendrimers as potential agents against fibrillation of a-synuclein, a Parkinson's disease-related protein. Macromol. Biosci. 9, 230–238.

Sardar Sinha, M., Villamil Giraldo, A.M., Ollinger, K., Hallbeck, M., Civitelli, L., 2018. Lipid vesicles affect the aggregation of 4-hydroxy-2-nonenal-modified α-synuclein oligomers. Biochim. Biophys. Acta (BBA) - Mol. Basis Dis. 1864 (9 Pt B), 3060–3068.

Schmid, A.W., Fauvet, B., Moniatte, M., Lashuel, H.A., 2013. Alpha-synuclein post-translational modifications as potential biomarkers for Parkinson disease and other synucleinopathies. Mol. Cell. Proteom. 12, 3543–3558.

Schommer, J., Marwarha, G., Schommer, T., Flick, T., Lund, J., Ghribi, O., 2018. 27-Hydroxycholesterol increases α-synuclein protein levels through proteasomal inhibition in human dopaminergic neurons. Schommer et al. BMC Neurosci. 19, 17.

Semerdzhiev, S.A., Dekker, D.R., Subramaniam, V., Claessens, M.M., 2014. Self-assembly of protein fibrils into suprafibrillar aggregates: bridging the nano-and mesoscale. ACS Nano 8, 5543–5551.

Siddiqi, K.S., Husen, A., Sohrab, S.S., Yassin, M.O., 2018. Recent status of nanomaterial fabrication and their potential applications in neurological disease management. Nanoscale Research Letters 13, 231.

Simon, K.C., Chen, H., Schwarzschild, M., Ascherio, A., 2007. Hypertension, hypercholesterolemia, diabetes, and risk of Parkinson disease. Neurology 69, 1688–1695.

Sode, K., Ochiai, S., Kobayashi, N., Usuzaka, E., 2006. Effect of reparation of repeat sequences in the human alpha-synuclein on fibrillation ability. Int. J. Biol. Sci. 3, 1–7.

Vila, M., Vukosavic, S., Jackson-Lewis, V., Neystat, M., Jakowec, M., Przedborski, S., 2000. α-Synuclein up-regulation in substantia nigra dopaminergic neurons following administration of the parkinsonian toxin MPTP. J. Neurochem. 74, 721–729.

Wakabayashi, K., Tanji, K., Mori, F., Takahashi, H., 2007. The Lewy body in Parkinson's disease: molecules implicated in the formation and degradation of a-synuclein aggregates. Neuropathology 27, 494–506.

Webb, J.L., Ravikumar, B., Atkins, J., Skepper, J.N., Rubinsztein, D.C., 2003. α-Synuclein is degraded by both autophagy and the proteasome. J. Biol. Chem. 278, 25009–25013.

Xiang, W., Schlachetzki, J.C., Helling, S., Bussmann, J.C., Berlinghof, M., Schaffer, T.E., et al., 2013. Oxidative stress-induced posttranslational modifications of alpha-synuclein: specific modification of alpha-synuclein by 4-hydroxy-2-nonenal increases dopaminergic toxicity. Mol. Cell. Neurosci. 54, 71–83.

Yamin, G., Glaser, C.B., Uversky, V.N., Fink, A.L., 2003. Certain metals trigger fibrillation of methionine-oxidized alpha-synuclein. J. Biol. Chem. 278, 27630–27635.

Yang, J.A., Lin, W., Woods, W.S., George, J.M., Murphy, C.J., 2014. α-Synuclein's adsorption, conformation, and orientation on cationic gold nanoparticle surfaces seeds global conformation change. J. Phys. Chem. B 118, 3559–3571.

Yu, J., Lyubchenko, Y.L., 2009. Early stages for Parkinson's development: alpha synuclein misfolding and aggregation. J. Neuroimmune Pharmacol. 4, 10–16.

Zhang, N., Yan, F., Liang, X., Wu, M., Shen, Y., Chen, M., et al., 2018. Localized delivery of curcumin into brain with polysorbate 80-modified cerasomes by ultrasound-targeted microbubble destruction for improved Parkinson's disease therapy. Theranostics 8, 2264–2277.

Zharikov, A.D., Cannon, J.R., Tapias, V., Bai, Q., Horowitz, M.P., Shah, V., et al., 2015. shRNA targeting alpha-synuclein prevents neurodegeneration in a Parkinson's disease model. J. Clin. Investig. 125, 2721–2735.

Further reading

An, Y., Tang, L., Jiang, X., Chen, H., Yang, M., Jin, L., et al., 2010. A photoelectrochemical immunosensor based on Au-doped TiO2 nanotube arrays for the detection ofα-synuclein. Chemistry 16, 14439–14446.

Cohlberg, J.A., Li, J., Uversky, V.N., Fink, A.L., 2002. Heparin and other glycosaminoglycans stimulate the formation of amyloid fibrils from alpha-synuclein in vitro. Biochemistry 41, 1502–1511.

Giehm, L., Oliveira, C.L., Christiansen, G., Pedersen, J.S., Otzen, D.E., 2010. SDS-induced fibrillation of alpha-synuclein: an alternative fibrillation pathway. J. Mol. Biol. 401, 115–133.

Monte, D.A., 2010. Alpha-synuclein suppression by targeted small interfering RNA in the primate substantia nigra. PLoS One 5, e12122.

CHAPTER 10

Pros and cons of Parkinson's disease therapeutics

1. Introduction

Parkinson's disease (PD) is among the most common neurodegenerative disorder characterized by degeneration of dopamine (DA)-producing neurons (Soursou et al., 2015). PD symptoms include bradykinesia, tremors, postural instability, and gait abnormality (Mayeux, 2003; Soursou et al., 2015). Despite the availability of abundant noninvasive procedures and combined tests for PD diagnosis, prognosis of early-onset PD is difficult (Soursou et al., 2015). Several lines of evidence have shown that early and long-term treatment with nonsteroidal antiinflammatory drugs protects neurons in neurodegenerative diseases, but the similar treatment after the onset is unsuccessful (Lambert et al., 2013; Carpanini et al., 2019). Complement system plays an important role in homeostasis, development, and regeneration in the central nervous system (CNS). Nevertheless, complement destabilization results in significant damage and disease (Carpanini et al., 2019). Therefore, complement targeting has long been in use for multifarious CNS diseases (Carpanini et al., 2019).

2. α-Synuclein

Accumulation of wild-type α-synuclein has long been implicated in a wide range of neurodegenerative diseases (Spillantini et al., 1997, 1998; Gai et al., 1998; Arawaka et al., 1998; Trojanowski et al., 1998; Takeda et al., 1998; Wakabayashi et al., 1997, 1998; Lucking and Brice, 2000; Daniel et al., 2019). α-Synuclein aggregation attenuators used currently in the PD treatment include multifarious phenols, flavonoids, catechols, rifampicin, anle 138b (Li et al., 2004; Ehrnhoefer et al., 2008; Ono and Yamada, 2006; Masuda et al., 2006; Meng et al., 2009, 2010; Di Giovanni et al., 2010; Wagner et al., 2013; Lorenzen et al., 2014; Deeg et al., 2015; Fernandes

Parkinson's Disease Therapeutics
ISBN 978-0-12-819882-7
https://doi.org/10.1016/B978-0-12-819882-7.00010-6

et al., 2017; Daniel et al., 2019). Oxidation-based nordihydroguaiaretic acid cyclization promoted its interaction with α-synuclein and substantially prevented the fibrillation process in *Caenorhabditis elegans* (Daniel et al., 2019).

3. *Caenorhabditis elegans*: a potential animal model for Parkinson's disease

Of the numerous PD models, the nematode round worm *C. elegans* has been corroborated as the substantial one with noteworthy replication of disease characteristics (Gaeta et al., 2019). These worms have DA neurons and facilitate study of genes, mechanisms, and other factors entailed in the DA neuron degeneration, which indeed is difficult to explore in complex organisms such as mammals (Gaeta et al., 2019). Initial studies in these worms followed by their phenotypes and further validation in complex organisms provide a fast-track to discover PD treatments in humans (Harrington et al., 2010; Dexter et al., 2012; Martinez et al., 2017; Cooper and Raamsdonk, 2018; Gaeta et al., 2019). A recent study unearthed the genes involved in enhancement or deterioration of the sensitivity to histone deacetylase inhibitor and valproic acid (VA) through a large RNAi synthetic lethality screen (Forthun et al., 2012; Gaeta et al., 2019). VA substantially defends DA neurons from degeneration in *C. elegans* model of PD. Variations in the chromatin state probably govern whether neurons are vulnerable or robust to α-synuclein toxicity (Kautu et al., 2013; Gaeta et al., 2019).

4. Blood–brain barrier

Human brain comprises approximately 100 billion neurons and is protected by intact blood–brain barrier (BBB). It has the smallest capillaries of 7–10 μm (Wong et al., 2013, Saeedi et al., 2019). It has no intracellular and valvar gaps, thus reducing the permeability significantly (Pardridge, 2002; Kas, 2004; Silva, 2008; Kabanov and Batrakova, 2004, Saeedi et al., 2019). To accomplish brain targeting through BBB in diseases with stable BBB, invasive and noninvasive techniques have been extensively used (Lu et al., 2014; Kasinathan et al., 2015; Carpanini et al., 2019). Using focused ultrasound waves, BBB has been temporarily impaired to localize "osmotic shock" agents (van Tellingen et al., 2015; Carpanini et al., 2019). Trojan horse delivery, chimeric peptides, nanoparticles (NPs), and viral vectors are a few hitherto employed brain drug targeting strategies (Carpanini et al., 2019).

5. Stem cell therapy

Stem cell therapy has been used to overcome AD and PD symptoms currently (Rosi and Cattaneo, 2002; Gogel et al., 2011; Abugable et al., 2017, Fleifel et al., 2018). Since the introduction of neural grafting in 1987 by Madrazo, it has been accepted as a potential therapeutic avenue in PD treatment (Fleifel et al., 2018). Two of the NIH-funded double blind placebo control trials in which PD patients were grafted with human fetal ventral mesencephalic cells unfortunately failed to improve PD symptoms and additionally induced graft-induced dyskinesia (Krack et al., 2000; Freed et al., 2001; Hagell et al., 2002; Olanow et al., 2003; Ma et al., 2010, Fleifel et al., 2018). To improve these studies, human embryonic stem cells were used in 1998 to produce dopaminergic neurons in both in vitro (Thomson, 1998; Kawasaki et al., 2000; Kim et al., 2002) and in vivo (Perrier et al., 2004; Yan et al., 2005; Park et al., 2005, Fleifel et al., 2018). However, they could not produce adequate midbrain neurons and could render limited amelioration, further inducing tumor formation in inadequately differentiated cells (Roy et al., 2006; Sonntag et al., 2007, Fleifel et al., 2018). Later in 2007 and 2008, it was understood that unlike all other neurons which are obtained from neuroepithelial cells, DA neurons are obtained from different source of cells (Ono et al., 2007; Bonilla et al., 2008, Fleifel et al., 2018). To develop appropriate DA neurons, floor plate—derived cells expressing appropriate markers of midbrain DA neurons were used, which successfully ameliorated motor functions in rodent models of PD-like human fetal DA neurons (Fasano et al., 2010; Kriks et al., 2011; Steinbeck et al., 2015; Kirkeby et al., 2012; Grealish et al., 2014, Fleifel et al., 2018). Chinese researchers introduced embryonic stem cells into PD patients, which produced mature DA-producing neurons (Cyranoski, 2017, Fleifel et al., 2018). Furthermore, Dr. Andrew Evans, Royal Melbourne Hospital neurologist, and his team accomplished world-first neural stem cell transplant (Fleifel et al., 2018).

6. Influence of gut microbiota on neurodegenerative diseases

Interestingly, metabolites and small molecular products generated by gut flora profoundly influence chemical signaling pathway and establish cometabolism between the gut flora and the brain, which is called the microbiota—gut—brain (MGB) axis. The perturbance of MGB axis deteriorates mental health and augments severity of neurodegenerative diseases such as AD, PD, anxiety, depression, and plethora of mental diseases (Obrenovich, 2017, 2018, Jones et al., 2019). The genetic kill switch, which can specifically kill harmful members of gut flora without affecting chemical signaling between gut flora and the brain, has been extensively used currently (Obrenovich et al., 2018, Jones et al., 2019).

Neurodegenerative disorders such as AD and PD are linked with dysbiosis of microbiota in a symbiotic relationship of gut—brain axis (Ji et al., 2014, Jones et al., 2019). Microbiota plays an essential role in generating signal molecules and hormones from intestinal metabolism that influences neurological disease and CNS inflammation (Jones et al., 2019). Dysregulation of microbiome promotes pathogenesis of neurodegenerative diseases (Ghaisas et al., 2016; Zuberi et al., 2017, Jones et al., 2019). In line with this, microbiota has been considered the substantial therapeutic target in the treatment of AD and PD (Jones et al., 2019).

6.1 Nanotechnology

Nanotechnological studies have been targeted at neurodegenerative disorders such as AD and PD currently (Vlamos and Alexiou, 2015a,b,c; Soursou et al., 2015). It has been shown that nanomaterials potentially enhance neurodegenerative disease therapeutics, in turn reducing the side effects significantly (Soursou et al., 2015). Multifarious drug delivery systems (DDS) employed to overcome BBB include polymeric NPs, nanospheres, and nanocapsules, which show higher drug loading capacity and low systemic toxicity (Muller and Keck, 2004; Modi et al., 2010; Soursou et al., 2015). Polymer NPs have been found to ferry BBB successfully and aid the brain drug delivery significantly (Amoozgar and Yeo, 2012; Neha et al., 2013; Kreuter, 2014, Saeedi et al., 2019). NPs ferry through the tight junctions of endothelial cells of the vessels and facilitate the drug passage through BBB (Saeedi et al., 2019). Endocytosis and transcytosis are other routes of drug transport through endothelial cell layer (Fischer et al., 2014; Herda et al., 2014, Saeedi et al., 2019). Moreover, NPs can target particular cells when tethered or coated with ligands and by attachment of specific ligands they can pass through the BBB by receptor-mediated transcytosis (Fillebeen et al., 1999; Ueno et al., 2010; Chen and Liu, 2012, Saeedi et al., 2019) (see Table 10.1).

6.2 Lactoferrin-functionalized nanoparticles

Although BBB is a robust impediment in PD therapy, yet gene therapy is considered the most promising (Soursou et al., 2015). Although lactoferrin (Lf)-functionalized NPs can effectively span BBB and target brain, Huang

Table 10.1 List of multifarious therapeutics, therapeutic effect, and references.

Name of the therapeutic	Therapeutic effect	Reference
Nonsteroidal antiinflammatory drugs	Protect neurons in neurodegenerative diseases	Lambert et al. (2013), Carpanini et al. (2019)
Phenols, flavonoids, catechols, rifampicin, anle 138b	Inhibition of α-synuclein aggregation	Li et al. (2004), Ehrnhoefer et al. (2008), Ono and Yamada (2006), Masuda et al. (2006), Meng et al. (2009), 2010, Di Giovanni et al. (2010), Wagner et al. (2013), Lorenzen et al. (2014), Deeg et al. (2015), Fernandes et al. (2017), Daniel et al. (2019)
Nordihydroguaiaretic acid	Prevented the fibrillation process in *Caenorhabditis elegans*	Daniel et al. (2019)
Valproic acid	Defend dopamine neurons from degeneration in *C. elegans* model of Parkinson's disease (PD)	Kautu et al. (2013), Gaeta et al. (2019)
Neural grafting	Potential therapeutic avenue in PD treatment	Fleifel et al., 2018
Nanoparticles (NPs)	**Therapeutic effect**	**Reference**
Polymeric NPs, nanospheres, and nanocapsules	Higher drug loading capacity and low systemic toxicity	Muller and Keck (2004), Modi et al. (2010), Soursou et al. (2015)
Polymer NPs	Ferry blood—brain barrier (BBB) successfully and aid the brain drug delivery	Amoozgar and Yeo (2012), Neha et al. (2013), Kreuter (2014), Saeedi et al., 2019
Lactoferrin (Lf)-functionalized NPs	Effectively span BBB and target brain	Hu et al. (2011), Soursou et al. (2015)
Rasagiline-loaded chitosan-coated poly(lactide-co-glycolide) NPs	Ameliorated PD symptoms after intranasal administration in Wistar rats	Niyaz (2017), Teleanu et al. (2019)
DNA plasmid-loaded NPs	Curtailed neurodegeneration	Kaplitt et al. (2007), Yurek et al., 2009a,b, Soursou et al. (2015)
Liposomal drug delivery systems for levodopa	Enhance the levodopa transfer through BBB	Saeedi et al., 2019

et al. (2009) reported them as substantial nonviral gene vector (Hu et al., 2011; Soursou et al., 2015). Frequent administrations of Lf-functionalized NPs like Lf-tethered polyethylene glycol—poly(lactide-co-glycolide) (PEG-PLGA) NPs exhibit common toxicity levels as DDS in the brain and can ameliorate PD symptoms such as dopaminergic neuronal loss despite the consequences in long-term noninvasive gene therapy (Hu et al., 2011; Soursou et al., 2015).

6.3 Nanotechnology for stem cell therapy

Recent studies have shown the efficacy of stem cells in neuroprotection and repair of degenerated neurons. They were designed with scaffold of polymer-based biodegradable nanofibers, which can be obtained by injection of an electrospinning, customized nanofiber into the scaffold (Nisbet et al., 2007; Sarojini et al., 2010; Soursou et al., 2015). In line with this, stem cells were produced in DA-generating nerve cells in chicken and mouse models despite the 20% unrelated stem cell production (Lindvall and Hagell, 2002; Sarojini et al., 2010; Patil and Rojekar, 2011; Soursou et al., 2015).

6.4 Rasagiline-loaded chitosan nanoparticles

Rasagiline (RSG) is a potential inhibitor of monoamine oxidase type B enzyme that attenuates biogenic amines such as DA in the CNS (Chen et al., 2007; Teleanu et al., 2019). Accordingly, to enhance the brain concentration of RSG, RSG-loaded chitosan-coated PLGA NPs were designed, which successfully ameliorated PD symptoms after intranasal administration in Wistar rats (Niyaz, 2017; Teleanu et al., 2019). NPs with compressed DNA plasmid loaded into it were delivered into the brain of animal models of PD and advanced PD patients successfully curtailed neurodegeneration (Kaplitt et al., 2007; Yurek et al., 2009a,b; Soursou et al., 2015).

6.5 Dopamine transporting nanoparticles

As DA cannot trespass BBB, it can be transported by specific metabolism and release strategies (Simpkins and Bodor, 1994; Soursou et al., 2015). Numerous biosensors have been used in the PD diagnosis for sensing DA such as carbon nanotubes and nanowires that are well accepted by human body (Soursou et al., 2015). These nanochips are multifunctional and light-weighted, with substantial mechanical strength (Soursou et al., 2015).

Besides their role in DA release and dopaminergic neuron regeneration, they interact with other biosensors introduced in the tremor producing area of the body (Soursou et al., 2015). Gurturk et al. (2017) developed liposomal DDS to enhance the levodopa transfer through BBB (Saeedi et al., 2019). In this formulation, glutathione, an antioxidant, acts as a cofactor and surface is conjugated with maltodextrin (Saeedi et al., 2019).

7. Conclusions and future perspectives

While the stem cell therapy is promising, there are a few challenges associated with it such as isolating and identifying appropriate stem cells from patient's tissues (Avinash et al., 2017, Fleifel et al., 2018). Additionally, during the expansion and passaging of stem cells, there is a possibility of phenotype getting disturbed and influencing the heterogenous group of cells produced making the genetic manipulation a difficult task (Zonari et al., 2017, Fleifel et al., 2018). Immune rejection is another challenge which is exerted by host against transplanted stem cells (Zhao et al., 2015, Fleifel et al., 2018). The primary reason for potentiating the *C. elegans* as substantial PD animal model is the ability of swift translation of results (Jones et al., 2019). In addition, this model facilitates the testing of therapeutic compounds on huge quantities of worms in a remarkably shorter period and cost-effective manner (Jones et al., 2019). Despite the lack of obvious therapeutic effect of multifarious hitherto tested compounds in *C. elegans*, there is a possibility to develop innovative compounds based on current therapeutics and their efficacy (Jones et al., 2019).

Despite the availability of numerous drug and neurotoxin-initiated PD models, appropriate replication of human PD has not been accomplished until now (Antony et al., 2011; Carpanini et al., 2019). There is an increasing need to develop novel therapeutic avenues for neurodegenerative diseases such as AD and PD because currently available therapies render insufficient therapeutic effect (Carpanini et al., 2019). Despite the development of innumerable novel drugs, there has been scant information regarding CNS drug targets (Morgan and Harris, 2015; Carpanini et al., 2019). Therefore, there is an augmenting need to quest the novel therapeutics by sequestering the expertise from multifarious specialties such as medicine, chemistry, and nanotechnology and combat dreadful neurodegenerative disorders.

References

Abugable, A.A., Awwad, D.A., Fleifel, D., Ali, M.M., El-Khamisy, S., Elserafy, M., 2017. Personalised Medicine: Genome Maintenance Lessons Learned from Studies in Yeast as a Model Organism, vol. 1007, pp. 157–178.

Amoozgar, Z., Yeo, Y., 2012. Recent advances in stealth coating of nanoparticle drug delivery systems. Wiley Interdiscip. Rev. Nanomed. Nanobiotechnol. 4, 219–233.

Antony, P.M., Diederich, N.J., Balling, R., 2011. Parkinson's disease mouse models in translational research. Mamm. Genome 22, 401–419.

Arawaka, S., Saito, Y., Murayama, S., Mori, H., 1998. Lewy body in neurodegeneration with brain iron accumulation type 1 is immunoreactive for alpha-synuclein. Neurology 51, 887–889.

Avinash, K., Malaippan, S., Dooraiswamy, J.N., 2017. Methods of isolation and characterization of stem cells from different regions of oral cavity using markers: a systematic review. Int. J. Stem Cells 10, 12–20.

Bonilla, S., Hall, A.C., Pinto, L., Attardo, A., Gotz, M., Huttner, W.B., et al., 2008. Identification of midbrain floor plate radial glia-like cells as dopaminergic progenitors. Glia 56, 809–820.

Carpanini, S.M., Torvell, M., Morgan, B.P., 2019. Therapeutic inhibition of the complement system in diseases of the central nervous system. Front. Immunol. 10, 362.

Chen, Y., Liu, L., 2012. Modern methods for delivery of drugs across the blood–brain barrier. Adv. Drug Deliv. Rev. 64, 640–665.

Chen, J.J., Swope, D.M., Dashtipour, K., 2007. Comprehensive review of rasagiline, a second-generation monoamine oxidase inhibitor, for the treatment of Parkinson's disease. Clin. Ther. 29, 1825–1849.

Cooper, J.F., Raamsdonk, M.V., 2018. Modeling Parkinson's disease in *C. elegans*. J. Parkinson's Dis. 8, 17–32.

Cyranoski, D., 2017. Trials of embryonic stem cells to launch in China. Nature 546, 15–16.

Daniels, M.J., Nourse Jr., J.B., Kim, H., Sainati, V., Schiavina, M., Murrali, M.G., et al., 2019. Cyclized NDGA modifies dynamic α-synuclein monomers preventing aggregation and toxicity. Sci. Rep. 9, 2937.

Dexter, P.M., Caldwell, K.A., Caldwell, G.A., 2012. A PredictableWorm: application of *Caenorhabditis elegans* for mechanistic investigation of movement disorders. Neurotherapeutics 9, 393–404.

Di Giovanni, S., Eleuteri, S., Paleologou, K.E., Yin, G., Zweckstetter, M., Carrupt, P.A., et al., 2010. Entacapone and tolcapone, two catechol O-methyltransferase inhibitors, block fibril formation of α-synuclein and β-amyloid and protect against amyloid-induced toxicity. J. Biol. Chem. 14941–14954.

Deeg, A.A., Reiner, A.M., Schmidt, F., Schueder, F., Ryazanov, S., Ruf, V.C., et al., 2015. Anle138b and related compounds are aggregation specific fluorescence markers and reveal high affinity binding to α-synuclein aggregates. Biochim. Biophys. Acta 1850, 1884–1890.

Ehrnhoefer, D.E., Bieschke, J., Boeddrich, A., Herbst, M., Masino, L., Lurz, R., et al., 2008. EGCG redirects amyloidogenic polypeptides into unstructured, off-pathway oligomers. Nat. Struct. Mol. Biol. 15, 558–566.

Fasano, C.A., Chambers, S.M., Lee, G., Tomishima, M.J., Studer, L., 2010. Efficient derivation of functional floor plate tissue from human embryonic stem cells. Cell Stem Cell 6, 336–347.

Fernandes, L., Moraes, N., Sagrillo, F.S., Magalhaes, A.V., de Moraes, M.C., Romao, L., et al., 2017. An ortho-iminoquinone compound reacts with lysine inhibiting aggregation while remodeling mature amyloid fibrils. ACS Chem. Neurosci. 8, 1704–1712.

Fillebeen, C., Descamps, L., Dehouck, M.P., Fenart, L., Benaissa, M., Spik, G., et al., 1999. Receptor-mediated transcytosis of lactoferrin through the blood-brain barrier. J. Biol. Chem. 274, 7011–7017.

Fischer, N.O., Weilhammer, D.R., Dunkle, A., Thomas, C., Hwang, M., Corzett, M., et al., 2014. Evaluation of nanolipoprotein particles (NLPs) as an in vivo delivery platform. PLoS One 9, e93342.

Fleifel, D., Rahmoon, M.A., AlOkda, A., Nasr, M., Elserafy, M., El-Khamisy, S.F., 2018. Recent advances in stem cells therapy: A focus on cancer, Parkinson's and Alzheimer's. J Genet Eng Biotechnol 16, 427–432.

Forthun, R.B., SenGupta, T., Skjeldam, H.K., Lindvall, J.M., McCormack, E., Gjertsen, B.T., et al., 2012. Cross-species functional genomic analysis identifies resistance genes of the histone deacetylase inhibitor valproic acid. PLoS One 7, e48992.

Freed, C.R., Greene, P.E., Breeze, R.E., Tsai, W.Y., DuMouchel, W., Kao, R., et al., 2001. Transplantation of embryonic dopamine neurons for severe Parkinson's disease. N. Engl. J. Med. 344, 710–719.

Gaeta, A.L., Caldwell, K.A., Caldwell, G.A., 2019. Found in translation: the utility of *C. elegans* alpha-synuclein models of Parkinson's disease. Brain Sci. 9, E73 pii.

Gai, W.P., Power, J.H., Blumbergs, P.C., Blessing, W.W., 1998. Multiple-system atrophy: a new alpha-synuclein disease? Lancet 352, 547–548.

Ghaisas, S., Maher, J., Kanthasamy, A., 2016. Gut microbiome in health and disease: linking the microbiome–gut–brain axis and environmental factors in the pathogenesis of systemic and neurodegenerative diseases. Pharmacol. Ther. 158, 52–62.

Gogel, S., Gubernator, M., Minger, S.L., 2011. Progress and prospects: stem cells and neurological diseases. Gene Ther. 18, 1–6.

Grealish, S., Diguet, E., Kirkeby, A., Mattsson, B., Heuer, A., Bramoulle, Y., et al., 2014. Human ESC-derived dopamine neurons show similar preclinical efficacy and potency to fetal neurons when grafted in a rat model of Parkinson's disease. Cell Stem Cell 15, 653–665.

Gurturk, Z., Tezcaner, A., Dalgic, A.D., Korkmaz, S., Keskin, D., 2017. Maltodextrin modified liposomes for drug delivery through the blood–brain barrier. Med Chem Comm 8, 1337–1345.

Harrington, A.J., Hamamichi, S., Caldwell, G.A., Caldwell, K.A., 2010. *C. elegans* as a model organism to investigate molecular pathways involved with Parkinson's disease. Dev. Dynam. 239, 1282–1295.

Hagell, P., Piccini, P., Bjorklund, A., Brundin, P., Rehncrona, S., Widner, H., 2002. Dyskinesias following neural transplantation in Parkinson's disease. Nat. Neurosci. 5, 627–628.

Herda, L.M., Polo, E., Kelly, P.M., Rocks, L., Hudecz, D., Dawson, K.A., 2014. Designing the future of nanomedicine: current barriers to targeted brain therapeutics. Eur. J. Nanomed. 6, 127–139.

Huang, R., Ke, W., Liu, Y., Wu, D., Feng, L., Jiang, C., et al., 2009. Gene therapy using lactoferrin-modified nanoparticles in a rotenone-induced chronic Parkinson model. J. Neurol. Sci. 290, 123–130.

Hu, K., Shi, Y., Jiang, W., Han, J., Huang, S., Jiang, X., 2011. Lactoferrin conjugated PEG-PLGA nanoparticles for brain delivery, Preparation, characterization and efficacy in Parkinson's disease. Int. J. Pharm. 415, 273–283.

Ji, W., Lee, D., Wong, E., Dadlani, P., Dinh, D., Huang, V., et al., 2014. Specific gene repression by CRISPRi system transferred through bacterial conjugation. ACS Synth. Biol. 3, 929–931.

Jones, L., Kumar, J., Mistry, A., Sankar Chittoor Mana, T., Perry, G., Reddy, V.P., 2019. The Transformative Possibilities of the Microbiota and Mycobiota for Health, Disease, Aging, and Technological Innovation. Biomedicines 7 (2).

Kabanov, A., Batrakova, E., 2004. New technologies for drug delivery across the blood brain barrier. Curr. Pharmaceut. Des. 10, 1355–1363.

Kaplitt, M.G., Feigin, A., Tang, C., Fitzsimons, H.L., Mattis, P., Lawlor, P.A., et al., 2007. Safety and tolerability of gene therapy with an adeno associated virus (AAV) borne GAD gene for Parkinson's disease, an open label, phase I trial. Lancet 369, 2097–2105.

Kasinathan, N., Jagani, H.V., Alex, A.T., Volety, S.M., Rao, J.V., 2015. Strategies for drug delivery to the central nervous system by systemic route. Drug Deliv. 22, 243–257.

Kas, H.S., 2004. Drug Delivery to Brain by Microparticulate Systems, Biomaterials. Springer, pp. 221–230.

Kreuter, J., 2014. Drug delivery to the central nervous system by polymeric nanoparticles: what do we know? Adv. Drug Deliv. Rev. 71, 2–14.

Kautu, B.B., Carrasquilla, A., Hicks, M.L., Caldwell, K.A., Caldwell, G.A., 2013. Valproic acid ameliorates *C. elegans* dopaminergic neurodegeneration with implications for ERK-MAPK signaling. Neurosci. Lett. 541, 116–119.

Kawasaki, H., Mizuseki, K., Nishikawa, S., Kaneko, S., Kuwana, Y., Nakanishi, S., et al., 2000. Induction of midbrain dopaminergic neurons from ES cells by stromal cell-derived inducing activity. Neuron 28, 31–40.

Kim, J.H., Auerbach, J.M., Rodriguez-Gomez, J.A., Velasco, I., Gavin, D., Lumelsky, N., et al., 2002. Dopamine neurons derived from embryonic stem cells function in an animal model of Parkinson's disease. Nature 418, 50–56.

Kirkeby, A., Grealish, S., Wolf, D.A., Nelander, J., Wood, J., Lundblad, M., et al., 2012. Generation of regionally specified neural progenitors and functional neurons from human embryonic stem cells under defined conditions. Cell Rep. 1, 703–714.

Kriks, S., Shim, J.W., Piao, J., Ganat, Y.M., Wakeman, D.R., Xie, Z., et al., 2011. Dopamine neurons derived from human ES cells efficiently engraft in animal models of Parkinson's disease. Nature 480, 547–551.

Krack, P., Poepping, M., Weinert, D., Schrader, B., Deuschl, G., 2000. Thalamic, pallidal, or subthalamic surgery for Parkinson's disease? J. Neurol. 247, 122–134.

Lambert, J.C., Ibrahim-Verbaas, C.A., Harold, D., Naj, A.C., Sims, R., Bellenguez, C., et al., 2013. Meta-analysis of 74,046 individuals identifies 11 new susceptibility loci for Alzheimer's disease. Nat. Genet. 45, 1452.

Li, J., Zhu, M., Rajamani, S., Uversky, V.N., Fink, A.L., 2004. Rifampicin inhibits alpha-synuclein fibrillation and disaggregates fibrils. Chem. Biol. 11, 1513–1521.

Lindvall, O., Hagell, P., 2002. Role of cell therapy in Parkinson's disease. Neurosurg. Focus 13, e2.

Lorenzen, N., Nielsen, S.B., Yoshimura, Y., Vad, B.S., Andersen, C.B., Betzer, C., et al., 2014. How epigallocatechin gallate can inhibit α-synuclein oligomer toxicity *in vitro*. J. Biol. Chem. 289, 21299–21310.

Lu, C.T., Zhao, Y.Z., Wong, H.L., Cai, J., Peng, L., Tian, X.Q., 2014. Current approaches to enhance CNS delivery of drugs across the brain barriers. Int. J. Nanomed. 9, 2241–2257.

Lucking, C.B., Brice, A., 2000. Alpha-synuclein and Parkinson's disease. Cell. Mol. Life Sci. 57, 1894–1908.

Ma, Y., Tang, C., Chaly, T., Greene, P., Breeze, R., Fahn, S., et al., 2010. Dopamine cell implantation in Parkinson's disease: long-term clinical and 18F-FDOPA PET outcomes. J. Nucl. Med. 51, 7–15.

Mayeux, R., 2003. Epidemiology of neurodegeneration. Annu. Rev. Neurosci. 26, 81–104.

Martinez, B.A., Caldwell, K.A., Caldwell, G.A., 2017. *C. elegans* as a model system to accelerate discovery for Parkinson disease. Curr. Opin. Genet. Dev. 44, 102–109.

Masuda, M., Suzuki, N., Taniguchi, S., Oikawa, T., Nonaka, T., Iwatsubo, T., et al., 2006. Small molecule inhibitors of α-synuclein filament assembly. Biochemistry 45, 6085–6094.

Meng, X., Munishkina, L.A., Fink, A.L., Uversky, V.N., 2010. Effects of various flavonoids on the α-synuclein fibrillation process. Parkinsons Dis 1–16.

Meng, X., Munishkina, L.A., Fink, A.L., Uversky, V.N., 2009. Molecular mechanisms underlying the flavonoid-induced inhibition of α-synuclein fibrillation. Biochemistry 48, 8206–8224.

Modi, G., Pillay, V., Choonara, Y.E., 2010. Advances in the treatment of neurodegenerative disorders employing nanotechnology. Ann. NY Acad. Sci. 1184, 154–172.

Morgan, B.P., Harris, C.L., 2015. Complement, a target for therapy in inflammatory and degenerative diseases. Nat. Rev. Drug Discov. 14, 857–877.

Muller, R.H., Keck, C.M., 2004. Drug delivery to the brain-realization by novel drug carriers. J. Nanosci. Nanotechnol. 4, 471–483.

Neha, B., Ganesh, B., Preeti, K., 2013. Drug delivery to the brain using polymeric nanoparticles: a review. Int. J. Pharm. Life Sci. 2, 107–132.

Nisbet, D.R., Crompton, K.E., Horne, M.K., Finkelstein, D.I., Forsythe, J.S., 2007. Neural tissue engineering of the CNS using hydro- gels. J. Biomed. Mater. Res. Part B App Biomat 87, 251–263.

Niyaz, A., 2017. Rasagiline-encapsulated chitosan-coated PLGA nanoparticles targeted to the brain in the treatment of Parkinson's disease. J. Liq. Chromatogr. Relat. Technol. 40, 677–690.

Obrenovich, M., 2018. Leaky Gut, Leaky Brain? Microorganisms, vol. 6, p. E107 pii.

Obrenovich, M., Rai, H., Mana, T.S., Shola, D., McCloskey, B., 2017. Dietary Co-metabolism within the microbiota-gut-brain-endocrine metabolic interactome. BAO Microbiol 2, 22.

Olanow, C.W., Goetz, C.G., Kordower, J.H., Stoessl, A.J., Sossi, V., Brin, M.F., et al., 2003. A double-blind controlled trial of bilateral fetal nigral transplantation in Parkinson's disease. Ann. Neurol. 54, 403–414.

Ono, K., Yamada, M., 2006. Antioxidant compounds have potent anti-fibrillogenic and fibril-destabilizing effects for alpha-synuclein fibrils *in vitro*. J. Neurochem. 97, 105–115.

Ono, Y., Nakatani, T., Sakamoto, Y., Mizuhara, E., Minaki, Y., Kumai, M., et al., 2007. Differences in neurogenic potential in floor plate cells along an anteroposterior location: midbrain dopaminergic neurons originate from mesencephalic floor plate cells. Development 134, 3213–3225.

Park, C.H., Minn, Y.K., Lee, J.Y., Choi, D.H., Chang, M.Y., Shim, J.W., et al., 2005. In vitro and in vivo analyses of human embryonic stem cell derived dopamine neurons. J. Neurochem. 92, 1265–1276.

Pardridge, W.M., 2002. Targeting neurotherapeutic agents through the blood-brain barrier. Arch. Neurol. 59, 35–40.

Patil, P., Rojekar, S., 2011. Self assembled cyclodextrin nanoparticles as drug carrier. Int. J. Pharm. Biol. Sci. 5, 569–588.

Perrier, A.L., Tabar, V., Barberi, T., Rubio, M.E., Bruses, J., Topf, N., et al., 2004. Derivation of midbrain dopamine neurons from human embryonic stem cells. Proc. Natl. Acad. Sci. U.S.A. 101, 12543–12548.

Rossi, F., Cattaneo, E., 2002. Neural stem cell therapy for neurological diseases: dreams and reality. Neuroscience 3, 401–409.

Roy, N.S., Cleren, C., Singh, S.K., Yang, L., Beal, M.F., Goldman, S.A., 2006. Functional engraftment of human ES cell-derived dopaminergic neurons enriched by coculture with telomerase-immortalized midbrain astrocytes. Nat. Med. 12, 1259–1268.

Saeedi, M., Eslamifar, M., Khezri, K., Dizaj, S.M., 2019. Applications of nanotechnology in drug delivery to the central nervous system. Biomed Pharmacother 111, 666–675.

Sarojini, S., Rajasekar, S., Kumaravelou, K., 2010. Carbon nanotubes, a new weapon in health care treatment. Int. J. Pharm. Biol. Sci. 1, 644–649.

Silva, G.A., 2008. Nanotechnology approaches to crossing the blood-brain barrier and drug delivery to the CNS. BMC Neurosci. 9, S4.

Simpkins, J.W., Bodor, N., 1994. The brain-targeted delivery of dopamine using a redox-based chemical delivery system. Adv. Drug Deliv. Rev. 14, 243–249.

Sonntag, K.C., Pruszak, J., Yoshizaki, T., van Arensbergen, J., Sanchez-Pernaute, R., Isacson, O., 2007. Enhanced yield of neuroepithelial precursors and midbrain-like dopaminergic neurons from human embryonic stem cells using the bone morphogenic protein antagonist noggin. Stem Cells 25, 411–418.

Soursou, G., Alexiou, A., Ashraf, G.M., Siyal, A.A., Mushtaq, G., Kamal, M.A., 2015. Applications of nanotechnology in diagnostics and therapeutics of alzheimer's and Parkinson'sDisease. Curr. Drug Metabol. 16, 705–712.

Spillantini, M.G., Schmidt, M.L., Lee, V.M., Trojanowski, J.Q., Jakes, R., Goedert, M., 1997. alpha-Synuclein in Lewy bodies. Nature 388, 839–840.

Spillantini, M.G., Crowther, R.A., Jakes, R., Hasegawa, M., Goedert, M., 1998. alpha-Synuclein in filamentous inclusions of Lewy bodies from Parkinson's disease and dementia with lewy bodies. Proc. Natl. Acad. Sci. U. S. A. 95, 6469–6473.

Steinbeck, J.A., Choi, S.J., Mrejeru, A., Ganat, Y., Deisseroth, K., Sulzer, D., et al., 2015. Optogenetics enables functional analysis of human embryonic stem cell-derived grafts in a Parkinson's disease model. Nat. Biotechnol. 33, 204–209.

Takeda, A., Mallory, M., Sundsmo, M., Honer, W., Hansen, L., Masliah, E., 1998. Abnormal accumulation of NACP/α-synuclein in neurodegenerative disorders. Am. J. Pathol. 152, 367–372.

Teleanu, D.M., Negut, I., Grumezescu, V., Grumezescu, A.M., Teleanu, R.I., 2019. Nanomaterials for drug delivery to the central nervous system. Nanomaterials 9, E371 pii.

Thomson, J.A., 1998. Embryonic stem cell lines derived from human blastocysts. Science 282, 1145–1147.

Trojanowski, J.Q., Goedert, M., Iwatsubo, T., Lee, V.M., 1998. Fatal attractions: abnormal protein aggregation and neuron death in Parkinson's disease and Lewy body dementia. Cell Death Differ. 5, 832–837.

Ueno, M., Nakagawa, T., Wu, B., Onodera, M., Huang, C.L., Kusaka, T., 2010. Transporters in the brain endothelial barrier. Curr. Med. Chem. 17, 1125–1138.

van Tellingen, O., Yetkin-Arik, B., de Gooijer, M.C., Wesseling, P., Wurdinger, T., de Vries, H.E., 2015. Overcoming the blood-brain tumor barrier for effective glioblastoma treatment. Drug Resist. Updates 19, 1–12.

Vlamos, P., Alexiou, A., 2015a. Neurodegeneration, springer series, advances in experimental medicine and biology. In: Vlamos, P., Alexiou, A. (Eds.), World Congress on Geriatrics and Neurodegenerative Disease Research, Content Level, vol. 822. Research, ISBN 978-3-319-08926-3.

Vlamos, P., Alexiou, A., 2015b. Geriatrics, springer series, advances in experimental medicine and biology. In: Vlamos, P., Alexiou, A. (Eds.), World Congress on Geriatrics and Neurodegenerative Disease Research, Content Level, vol. 821. Research, ISBN 978-3-319-08938-6.

Vlamos, P., Alexiou, A., 2015c. Computational biology and bioinformatics, springer series, advances in experimental medicine and biology. In: Vlamos, P., Alexiou, A. (Eds.), World Congress on Geriatrics and Neurodegenerative Disease Research, Content Level, vol. 820. Research, ISBN 978-3-319-09011-5.

Wakabayashi, K., Yoshimoto, M., Tsuji, S., Takahashi, H., 1998. Alpha-synuclein immunoreactivity in glial cytoplasmic inclusions in multiple system atrophy. Neurosci. Lett. 249, 180–182.

Wakabayashi, K., Matsumoto, K., Takayama, K., Yoshimoto, M., Takahashi, H., 1997. NACP, a presynaptic protein, immunoreactivity in Lewy bodies in Parkinson's disease. Neurosci. Lett. 239, 45–48.

Wagner, J., Ryazanov, S., Leonov, A., Levin, J., Shi, S., Schmidt, F., et al., 2013. Anle138b: a novel oligomer modulator for disease-modifying therapy of neurodegenerative diseases such as prion and Parkinson's disease. Acta Neuropathol. 125, 795–813.

Wong, A., Ye, M., Levy, A.F., Rothstein, J.D., Bergles, D.E., Searson, P.C., 2013. The blood-brain barrier: an engineering perspective. Front. Neuroeng. 6, 7.

Yan, Y., Yang, D., Zarnowska, E.D., Du, Z., Werbel, B., Valliere, C., et al., 2005. Directed differentiation of dopaminergic neuronal subtypes from human embryonic stem cells. Stem Cells 23, 781–790.

Yurek, D.M., Fletcher, A.M., Smith, G.M., Seroogy, K.B., Ziady, A.G., Molter, J., et al., 2009a. Long-term transgene expression in the central nervous system using DNA nanoparticles. Mol. Ther. 17, 641–650.

Yurek, D.M., Fletcher, A.M., Kowalczyk, T.H., Padegimas, L., Cooper, M.J., 2009b. Compacted DNA nanoparticle gene transfer of GDNF to the rat striatum enhances the survival of grafted fetal dopamine neurons. Cell Transplant. 18, 1183–1196.

Zonari, E., Desantis, G., Petrillo, C., Boccalatte, F.E., Lidonnici, M.R., Kajaste-Rudnitski, A., et al., 2017. Efficient ex vivo engineering and expansion of highly purified human hematopoietic stem and progenitor cell populations for gene therapy. Stem Cell Rep. 8, 977–990.

Zhao, T., Zhang, Z.N., Westenskow, P.D., Todorova, D., Hu, Z., Lin, T., et al., 2015. Humanized mice reveal differential immunogenicity of cells derived from autologous induced pluripotent stem cells. Cell Stem Cell 17, 353–359.

Zuberi, A., Misba, L., Khan, A.U., 2017. CRISPR Interference (CRISPRi) inhibition of luxS gene expression in *E. coli*: an approach to inhibit biofilm. Front. Cell Infect. Microbiol. 7, 214.

Index

Note: 'Page numbers followed by "f" indicate figures and "t" indicate tables'.

A

Adeno-associated virus (AAV), 3
Adenosine triphosphate (ATP), 39
Alginate, 3
Alzheimer's disease (AD), 1, 13, 25–26, 40, 89–91
Amelioration, 43–45
Anthocyanins, 43
Antiapoptosis, 54
Antioxidant enzymes, 43–45
Antisense single-stranded oligos (AntimiRs), 1–2
Apolipoprotein E (ApoE), 5
Apomorphine, 28–29

B

Baicalein, 42
BBB. *See* Blood-brain barrier (BBB)
Biomaterials, 13–23
Bisdemethoxycurcumin (BDMC), 53
Blood-brain barrier (BBB), 3, 5, 14, 28, 65, 75–76, 76f, 89–91, 116
- curcumin, 55
- dysfunction, 25–26
- flavonoids, 40
- gold nanoparticles, 29–30
- impairment
  - cerium oxide nanoparticles, 27–28
  - nanoparticles, 26–27
- multifarious pathways, 25–26, 26f
- quercetin nanoparticles, 30

Brain targeted liposomes, 78, 79t–80t

C

Caenorhabditis elegans, 116
Carbon nanotubes (CNTs), 4–5
Catalase, 39
Cerebrospinal fiuid (CSF) biomarkers, 26
Cerium oxide (CeO_2) nanoparticles, 27–28, 69
Chinese herbs, 40
Chitosan nanoparticles, 93–94
Coenzyme Q, 66
Coenzyme Q10 (CoQ10), 66–69
- delivery system, 67–68

Curcumin, 51, 81–82, 108
- antiapoptosis, 54
- antioxidant activity, 51–52
- bioavailability, 53
- delivery systems, 55–56, 57t–58t
- derivatives, 53, 57t–58t
- heat shock proteins (HSPs), 53–54
- 6-hydroxydopamine (6-OHDA), 52
- metal chelation, 54–55
- 1-methyl-4-phenyl-1,2,3,6-tetrahydropyridine (MPTP)-induced PD models, 52
- α-synuclein targeting, 52
- therapeutic effects, 57t–58t

D

Demethoxycurcumin (DMC), 53
Dendrimers, 108
Diferuloylmethane. *See* Curcumin
Dopamine nanoparticles
- cerium oxide (CeO_2) nanoparticles, 69
- electric signal transmission, 1
- redox-active nanoparticles (RNPs), 69

Drug delivery systems (DDS), 14, 55–56, 75–76

E

Electric signal transmission, 1
Electrosteric stealth (ESS) liposomes, 92
Epigallocatechin-3-gallate (EGCG), 16, 40–41, 44

F

Fatty acid-binding protein (FABP), 103–104
Fibrillation, 52
Fibroblast growth factor-20 (FGF-20), 82
Flavonoids, 40

G

Ginsenoside, 44
Glutathione, 44
Gold nanoparticles (AuNPs), 29–30
Gut microbiota, 117–121
Gypenosides (GP), 43

H

Heat shock proteins (HSPs), 53–54, 105–106
High-density lipoprotein nanoparticles, 5
6-Hydroxydopamine (6-OHDA), 52

I

Immunotherapy, 17

K

Kaempferol, 40

L

Lactoferrin-functionalized nanoparticles, 4–5, 92–93, 118–120
Levodopa (L-DOPA), 2, 15, 75
Lipophilicity, 2, 25–26
Liposomes, 5, 76–77, 92
  demerits, 83
  dopamine delivery, 78–81
  natural compounds, 81–82
  Parkinson's disease, 78

M

Malondialdehyde, 51–52
Manganese oxide NPs, 16
Metal chelation, 44–45, 54–55
Micro-ribonucleic acid (miRNA), 1–2
Mitochondrial DNA (mtDNA), 39
Mitochondrial membrane potential (MMP), 42–43
Multidrug resistance proteins (MRPs), 2
Multifarious nanoparticles, 4
Multispecific organic anion transporter (MOAT), 2

N

Nanoliposomes, 77
Nanoparticles (NPs), 14–15, 26–27, 29. *See also specific nanoparticles*
  blood-brain barrier (BBB), 28
  brain delivery, 3–4
  cerium oxide, 27–28
  drug delivery, 3–5
  gene delivery, 4–5
  high-density lipoprotein, 5
  multifarious, 4
  quercetin, 30, 31t–32t
  redox. *See* Redox nanoparticles
  silver and gold, 3–4, 17, 29–30
  tissue engineering, 3–4
Nanostructured lipid carriers (NLCs), 82–83, 94–95
Nanotechnology, 3–4, 67, 118, 119t
  demerits, 6
  Parkinson's disease diagnosis, 106, 107t
  Parkinson's disease therapy, 106–108
  shortcomings, 17
  stem cell therapy, 120
Nanozyme, 68
Naringenin, 41
Nasal drug delivery
  brain-targeted drugs and nanoparticles, 89, 90t
  challenges, 91–95
    chitosan nanoparticles, 93–94
    liposomes, 92
    nanostructured lipid carriers, 94–95
    nanotechnology, 91–92
    peptide-based carriers, 95
    polysaccharide-based nanoparticles, 95
    surface functionalization, 92–93
  nose to brain delivery, 95–96
Natural compounds
  antioxidant enzymes, amelioration of, 43–45
  delivery systems, 15–16
  liposomes, 81–82

oxidative stress, 39
plant-derived, 40–43
Neuroinfiammation, 40
Neurotensin (NTS) polyplex, 4–5
Neurturin (NRTN), 15
Nicotine, 44–45
NTS receptor type1 (NTSR1), 15

O

Oxidative stress, 13–14, 39

P

Peptide-based carriers, 95
P-glycoprotein (P-gp), 2, 25–26, 75–76
Piperine, 3
Plant-derived compounds
baicalein, 42
Chinese herbs, 40
flavonoids, 40
mitochondria, 40–41
naringenin, 41
polyphenolic fiavonoids, 43
polyphenols, 42–43
silibinin, 41
Polylactic-co-glycolic acid (PLGA), 68
Polyphenolic fiavonoids, 43
Polyphenols, 42–43
Posttranscriptional gene silencing, 1–2
Protein nitration, 13–14

Q

Quantum dots (QDs), 66
Quercetin nanoparticles, 30, 31t–32t

R

Rasagiline (RSG), 3, 120
Reactive oxygen species (ROS), 1, 39, 51–52, 65
Redox nanoparticles, 65–66, 69
coenzyme Q10 (CoQ10), 66–69
delivery system, 67–68
dopamine nanoparticles, 68–69
nanotechnology, 67
nanozyme, 68
Parkinson's disease diagnosis, 66
Resveratrol, 42
Retinoic acid (RA), 16
ROS. *See* Reactive oxygen species (ROS)
Rotigotine, 2, 28–29

S

Silibinin, 41
Silver nanoparticles (SNP), 17
Small hydrophobic molecules (SHM), 78–81
Solid lipid nanoparticles (SLNs), 77, 82–83
Stem cell therapy, 117
Stimuli-sensitive liposomes, 77
Superoxide dismutase (SOD), 39
Surface functionalization, 4
brain targeting, 77–83
nasal drug delivery, 92–93
α-Synuclein aggregation, 108–109
α-Synuclein targeted nanoparticles, 52, 103–104, 115–116
early diagnosis, 105
therapy, 105–109
fibrillation, 105
1-methyl-4-phenyl-1,2,3,6-tetrahydro-pyridine (MPTP) treatment, 104–105

T

Tarenfiurbil, 3

U

Ubiquinone, 66

V

Vaccination, 17
Viral vectors, 2f, 3–6
Vitamin B3, 44–45

Printed in the United States
By Bookmasters